An Introduction to Radio Frequency Engineering

This book provides a comprehensive introduction to radio frequency (RF) engineering, using a straightforward and easily understood approach combined with numerous worked examples, illustrations and homework problems. The author has focused on minimising the mathematics needed to grasp the subject while providing a solid theoretical foundation for the student. Emphasis is also placed on the practical aspects of radio engineering. The book provides a broad coverage of RF systems, circuit design, antennas, propagation and digital techniques. Written for upper-level undergraduate courses, it will also provide an excellent introduction to the subject for graduate students, researchers and practising engineers.

CHRISTOPHER COLEMAN is an associate professor of electrical and electronic engineering at the University of Adelaide. He has held research posts at Imperial College London and the University College of Wales, and faculty positions at the University of Nottingham, UK, and the University of Wollongong, Australia. From 1990 until 1999 he served as a principal research scientist on Australia's Jindalee over the horizon radar project. He is a member of the IEEE and an associate editor of the *IEEE Antennas and Wireless Propagation Letters*.

An Introduction to Radio Frequency Engineering

Christopher Coleman

Department of Electrical and Electronic Engineering,
Adelaide University

CAMBRIDGE
UNIVERSITY PRESS

PUBLISHED BY THE PRESS SYNDICATE OF THE UNIVERSITY OF CAMBRIDGE
The Pitt Building, Trumpington Street, Cambridge, United Kingdom

CAMBRIDGE UNIVERSITY PRESS
The Edinburgh Building, Cambridge CB2 2RU, UK
40 West 20th Street, New York, NY 10011–4211, USA
477 Williamstown Road, Port Melbourne, VIC 3207, Australia
Ruiz de Alarcón 13, 28014 Madrid, Spain
Dock House, The Waterfront, Cape Town 8001, South Africa

http://www.cambridge.org

First published 2004

Printed in the United Kingdom at the University Press, Cambridge

Typefaces Times 10.5/14 pt. and Helvetica   *System* LATEX 2_ε [TB]

A catalogue record for this book is available from the British Library

Library of Congress Cataloguing in Publication data

Coleman, Christopher, 1950–
An introduction to radio frequency engineering / Christopher Coleman.
        p.   cm.
Includes bibliographical references and index.
ISBN 0-521-83481-3
1. Radio circuits – Design and construction.   2. Radio – Equipment and supplies – Design and construction.
3. Radio frequency.   I. Title.
TK6560.C64   2004
621.384 – dc22   2003055893

ISBN 0 521 83481 3 hardback

This book is dedicated to the memory of Les France and Peter Hattam.

Contents

List of figures   *page* xi
Preface   xxi
Acknowledgements   xxii

1 Basic concepts   1

   1.1 Radio waves   1
   1.2 Noise   9
   1.3 Sensitivity and selectivity   13
   1.4 Non-linearity in RF systems   14
   1.5 Digital modulation   19
   1.6 Spread spectrum systems   20
   1.7 Cellular radio   23
   1.8 Radar systems   24

2 Frequency selective circuits and matching   28

   2.1 Series resonant circuits   28
   2.2 Parallel resonant circuits   33
   2.3 Inductive transformers   34
   2.4 Tuned transformers   36
   2.5 Capacitive transformers   37
   2.6 L-network matching   38
   2.7 π-and T-networks   39
   2.8 Matching examples   41
   2.9 Component reality   44

3 Active devices and amplifiers   49

   3.1 The semiconductor diode   49
   3.2 Bipolar junction transistors   50

3.3   The Miller effect and BJT amplifiers   57
3.4   Differential amplifiers   64
3.5   Feedback   67
3.6   Field-effect transistors   71
3.7   FET amplifiers   76
3.8   Amplifier noise   82

4   Mixers, modulators and demodulators   86

4.1   Diode mixers   86
4.2   Transistor mixers   89
4.3   Transconductance mixers   93
4.4   Amplitude modulation   95
4.5   Angle modulation   101
4.6   Gain and amplitude control   103
4.7   A simple receiver design   105

5   Oscillators and phase locked loops   108

5.1   Feedback   108
5.2   The Colpitts oscillator   109
5.3   Stability and phase noise in oscillators   117
5.4   Voltage controlled oscillators   120
5.5   Negative resistance approach to oscillators   121
5.6   Phase locked loops   123
5.7   Analysis of a phase locked loop   123
5.8   Phase locked loop components   125
5.9   Phase locked loop applications   127

6   Transmission lines and scattering matrices   132

6.1   The transmission line model   132
6.2   Time-harmonic variations   135
6.3   Real transmission lines   139
6.4   Impedance transformation   142
6.5   Reflection coefficients   145
6.6 S parameters   148
6.7 S parameter amplifier design   154
6.8   The measurement of S parameters   161

6.9   Some useful multiport networks   162
6.10   Reflection coefficient approach to microwave oscillators   166

7   Power amplifiers   171

7.1   Class A   171
7.2   Class B   173
7.3   Class C   176
7.4   Class E   178
7.5   A design example   179
7.6   Transmission line transformers   181

8   Filters   187

8.1   Filter characteristics   187
8.2   Low- and high-pass filters   191
8.3   Band-pass filters   194
8.4   Conversion of filters to microstrip form   196

9   Electromagnetic waves   204

9.1   Maxwell's equations   204
9.2   Power flow   206
9.3   Electromagnetic waves   206
9.4   Oblique incidence   215
9.5   Guided wave propagation   218
9.6   Wave sources   224

10   Antennas   228

10.1   Dipole antennas   228
10.2   Effective length and gain   235
10.3   The monopole antenna   238
10.4   Feeding an antenna   243
10.5   Array antennas   247
10.6   Travelling wave antennas   258
10.7   Aperture antennas   260
10.8   Patch antennas   261

11   Propagation

266

11.1	Reciprocity theorem	266
11.2	Some consequences of reciprocity	269
11.3	Line-of-sight propagation and reflections	271
11.4	Diffraction	275
11.5	Refraction	277
11.6	Ground wave propagation	286
11.7	Propagation by scattering	287

12   Digital techniques in radio

293

12.1	The processing of digitised signals	293
12.2	Analogue-to-digital conversion	298
12.3	Digital receivers	301
12.4	Direct digital synthesis	305

Index

307

Figures

1.1	Radiation of a radial pulse.	*page* 2
1.2	Electromagnetic field generated by a system of accelerating charge.	3
1.3	Development of oscillating dipole field.	4
1.4	A dipole antenna and its gain pattern.	5
1.5	Circuit model of a transmit system.	6
1.6	Conventions for effective length.	6
1.7	A dipole antenna used to collect energy from an electromagnetic wave.	7
1.8	Circuit model of receive system.	7
1.9	Reciprocity principle.	8
1.10	Transmit/receive system.	8
1.11	Noise sources.	9
1.12	Typical antenna noise temperatures for a dipole and a variety of sources.	10
1.13	Cascaded amplifiers.	12
1.14	Measurement of noise figure using matched sources at different temperatures.	13
1.15	Simple receiver architecture.	14
1.16	Superheterodyne receiver.	14
1.17	Set-up for measuring third-order intercept point.	17
1.18	Graph depicting the third-order intercept point and 1 dB compression point.	18
1.19	A direct conversion receiver together with the performance of its individual stages.	18
1.20	Formation of a spread baseband signal.	21
1.21	A simple DSSS transmitter.	22
1.22	A simple DSSS receiver.	22
1.23	A cellular radio system.	23
1.24	A general radar configuration.	25
1.25	Moving target.	26
2.1	Series resonant circuit.	29
2.2	Transient response.	30

2.3	Behaviour of transfer function around resonance.	31
2.4	Series resonant circuit with non-ideal source.	32
2.5	Notch filter example.	32
2.6	Parallel resonant circuit.	33
2.7	Transformer conventions.	34
2.8	Ideal transformer.	35
2.9	Non-ideal transformer.	35
2.10	Tuned transformer.	36
2.11	Model of tuned transformer.	36
2.12	Capacitive transformer.	37
2.13	Resonant capacitive transformer.	37
2.14	L matching network.	38
2.15	π-network.	40
2.16	Analysis of π-network.	40
2.17	T-network.	40
2.18	Analysis of a T-network.	41
2.19	L-network example.	42
2.20	π-network example.	42
2.21	Analysis of π-network example.	43
2.22	Final π-network.	43
2.23	Single layer solenoid and spiral inductor.	44
2.24	Toroidal inductor and realistic inductor model.	45
2.25	Toroidal transformer and circuit model.	46
2.26	Parallel plate and parallel wire capacitors.	47
2.27	High frequency circuit models of realistic resistors and capacitors.	47
3.1	Junction diode.	50
3.2	High frequency model of a diode.	50
3.3	Bipolar junction transistor (BJT).	51
3.4	Typical BJT characteristics.	51
3.5	BJT load line.	52
3.6	Typical bias circuit.	53
3.7	Alternative bias circuits for a BJT common-emitter amplifier.	54
3.8	A common-emitter amplifier.	54
3.9	Simple BJT models.	55
3.10	Small signal model of common-emitter amplifier.	55
3.11	High frequency BJT model.	56
3.12	Alternative bias circuits for a BJT common-emitter amplifier.	56
3.13	Feedback amplifier and Miller equivalent.	57
3.14	Common-emitter amplifier model.	58
3.15	Miller transformed model.	58
3.16	A more complete BJT model.	59

3.17 A common-base amplifier.   61

3.18 Common-base amplifier model.   61

3.19 The cascode amplifier.   62

3.20 The emitter follower amplifier.   63

3.21 Emitter follower amplifier model.   63

3.22 A BJT differential amplifier.   65

3.23 Differential amplifier with current source.   66

3.24 A current mirror ($I_0 = (V_{CC} - V_{BE})/R$ and $V_{BE} \approx 0.7\,\text{V}$).   66

3.25 Active loads.   67

3.26 Differential amplifier with active loads.   67

3.27 A high frequency amplifier.   68

3.28 A feedback system.   68

3.29 BJT amplifier with feedback.   69

3.30 The calculation of resistances for bandwidth calculations.   70

3.31 The n-channel junction field-effect transistor (JFET).   71

3.32 The n-channel enhancement MOSFET (nMOS).   72

3.33 The p-channel enhancement MOSFET (pMOS).   72

3.34 Typical FET characteristics.   73

3.35 Typical $I_D - V_{GS}$ curves for (a) a JFET, (b) an nMOS enhancement FET and (c) a pMOS enhancement FET.   73

3.36 Bias configuration and bias spread characteristics.   74

3.37 FET load line.   74

3.38 Alternative biasing for a JFET.   75

3.39 Low frequency FET model.   75

3.40 Common-source amplifier.   76

3.41 Model of an FET device at RF frequencies.   77

3.42 Common-gate amplifier.   79

3.43 FET cascode amplifier.   79

3.44 Dual-gate MOSFET amplifier.   80

3.45 Source follower amplifier.   81

3.46 A CMOS differential amplifier.   81

3.47 FET active loads.   82

3.48 Amplifier with noise model.   82

3.49 Circuit of d.c.-coupled amplifier for Question 4.   84

3.50 An FET model that is appropriate at very high frequencies.   84

4.1 Diode mixer.   87

4.2 Balanced mixer based on a diode pair.   88

4.3 Diode ring mixer.   88

4.4 A typical single FET mixer.   89

4.5 Dual-gate MOSFET mixer.   91

4.6 BJT mixer.   91

4.7	Alternative BJT mixer.	92
4.8	A basic transconductance mixer.	93
4.9	Gilbert cell mixer.	94
4.10	Simple AM transmitter.	95
4.11	Simple AM detector.	96
4.12	Synchronous detector.	96
4.13	Sideband structure.	97
4.14	Simple SSB transmitter.	97
4.15	Variable frequency SSB transmitter.	98
4.16	Phasing SSB generation.	98
4.17	Analysis of the phasing technique.	99
4.18	Phasing SSB demodulation.	100
4.19	An RC network 90° phase shifter.	100
4.20	Traditional SSB receiver.	100
4.21	Phase modulator.	102
4.22	Simple FM demodulation.	102
4.23	Detector frequency response.	103
4.24	Automatic gain control.	104
4.25	Filter and amplifier for AGC.	104
4.26	Limiting amplifier.	105
4.27	A simple direct conversion receiver.	105
4.28	Modern form of DC receiver.	106
5.1	General feedback system.	109
5.2	Colpitts oscillator.	110
5.3	Simple model of Colpitts oscillator.	110
5.4	FET Colpitts with bias circuits.	111
5.5	General Colpitts oscillator.	111
5.6	Clapp and Hartley feedback circuits.	112
5.7	FET differential oscillator.	112
5.8	The drain current for a Colpitts oscillator.	114
5.9	Alternative Colpitts oscillator.	114
5.10	Source follower amplifier.	115
5.11	Simplified Colpitts oscillator model.	115
5.12	An FET Colpitts oscillator that allows output to be taken from the drain.	116
5.13	Oscillator with series LCR feedback.	117
5.14	Phase noise.	118
5.15	Illustration of reciprocal mixing.	119
5.16	Crystal resonator model.	120
5.17	Crystal Colpitts oscillator.	120
5.18	A Colpitts VCO.	121

5.19 Gunn diode.   121
5.20 Oscillator based on the Gunn diode.   122
5.21 Negative resistance circuit based on an FET.   122
5.22 General phase locked loop.   123
5.23 Response of a phase locked loop to a step change in reference phase.   125
5.24 Phase comparator.   126
5.25 Loop filter.   126
5.26 Phase/frequency detector based on D flip-flops and a charge pump.   127
5.27 Signal levels in the phase/frequency detector.   128
5.28 A simple FET PLL.   129
5.29 Phase locked loop FM demodulator.   129
5.30 Phase locked loop AM detector.   129
5.31 Frequency synthesiser.   129
5.32 Frequency dividers based on D flip-flops.   130
6.1   Coaxial and twin parallel wire transmission lines.   133
6.2   Lumped component model of transmission line.   133
6.3   Propagation on a transmission line with resistive load.   135
6.4   Terminated transmission line.   136
6.5   The input impedance of a loaded line ($Z_{in} = R_{in} + X_{in}$).   138
6.6   Microstrip transmission line.   140
6.7   Realistic model of transmission line.   141
6.8   The short-circuited line.   142
6.9   The open-circuited line.   143
6.10 The $\lambda/4$ transformer.   143
6.11 Stub matching.   144
6.12 Replacing the stub by a lumped component.   144
6.13 Matching example.   145
6.14 Reflection coefficient description of a one-port network.   146
6.15 A complex tuned circuit and associated Smith chart.   147
6.16 Smith charts.   147
6.17 Source and load connected by a transmission line.   147
6.18 S matrix description of two-port network.   149
6.19 Z matrix description of two-port network.   149
6.20 Symmetric T-network example.   150
6.21 Configuration for the derivation of S parameters.   151
6.22 A simple two-port model of an FET.   152
6.23 General amplifier with source and load.   154
6.24 Transistor amplifier.   154
6.25 Smith chart representation of amplifier stability.   157
6.26 Final amplifier design.   160
6.27 A simple directional coupler.   161

6.28   Set-up for the measurement of S parameters.   162
6.29   Microstrip form of Wilkinson power divider.   162
6.30   Microstrip form of directional coupler.   163
6.31   Analysis of microstrip coupler in terms of even and odd modes.   164
6.32   A branch line quadrature hybrid.   165
6.33   A microwave mixer that is based on a 90° hybrid and diode pair.   165
6.34   Microstrip circulator with application.   166
6.35   Amplifier configuration.   167
6.36   An FET negative resistance oscillator.   167
6.37   Model of FET negative resistance generator.   168
6.38   Microstrip implementation of a negative resistance oscillator.   168
6.39   Circuit for Smith chart example in Question 5.   169
6.40   Circuits for two-port S matrix calculations in Question 6.   169
7.1   Basic power amplifier circuit.   172
7.2   Class A operation.   172
7.3   Class B operation.   174
7.4   Push-pull class B amplifier.   175
7.5   A bias circuit for class B operation.   176
7.6   Class C amplifier.   177
7.7   Class E amplifier.   179
7.8   Power amplifier example.   180
7.9   Design of an output matching network.   180
7.10   Input matching network.   181
7.11   Transmission line transformer model.   182
7.12   Twisted pair transmission line.   183
7.13   Practical transformer.   183
7.14   4:1 transformer.   184
7.15   9:1 transformer.   184
7.16   Power combiner.   184
7.17   A 1:1 BALUN.   185
8.1   Ideal filters.   188
8.2   A simple low-pass filter.   188
8.3   A simple high-pass filter.   188
8.4   Two-pole filter example.   189
8.5   Transducer gain for two-pole filter.   189
8.6   Typical low-pass filter characteristics.   190
8.7   Comparison of Butterworth and Chebyshev filters.   191
8.8   Low-pass Butterworth filter characteristics.   191
8.9   Low-pass filters for $K = 1$ and $K = 2$.   192
8.10   Low-pass filters for $K = 3$ and $K = 5$.   192
8.11   Alternative low-pass Butterworth configurations.   193

8.12	High-pass filters for $K = 1$ and $K = 2$.	194
8.13	High-pass filters for $K = 3$ and $K = 5$.	194
8.14	High-pass filter characteristics.	195
8.15	Band-pass filter derived from a low-pass filter.	195
8.16	Transformation of filter characteristics.	195
8.17	Circuit with three-element filter.	197
8.18	Correspondence between lumped components and transmission lines.	197
8.19	Basic three-element filter.	197
8.20	Some of Kuroda's identities.	198
8.21	Filter with transmission line extensions.	199
8.22	Filter after application of Kuroda identities.	199
8.23	Transmission line realisation.	199
8.24	Prototype lumped component filter.	199
8.25	Prototype transmission line filter.	200
8.26	Final transmission line filter.	200
8.27	Microstrip realisation of the filter.	201
8.28	Simple parallel microstrip filter.	201
8.29	Parallel microstrip filter.	202
9.1	A plane electromagnetic pulse.	208
9.2	The decomposition of linear polarisation into circularly polarised modes.	210
9.3	Dispersion of pulses.	211
9.4	Group speed for a simple two-component signal.	212
9.5	Reflection and transmission at a plane interface.	212
9.6	Transmission through a screen.	213
9.7	Attenuation in a current carrying conductor.	215
9.8	The skin effect on a wire conductor.	215
9.9	Oblique incidence of plane waves.	216
9.10	Coaxial waveguide.	218
9.11	Microstrip geometry and fields.	219
9.12	Hollow rectangular waveguide.	219
9.13	Rectangular cavities with loop and probe coupling.	222
9.14	Dominant modes in rectangular and circular hollow waveguides.	222
9.15	High Q bandstop microstrip formed with dielectric resonator.	223
9.16	High stability oscillator that uses a dielectric resonator.	223
9.17	Field zones in relation to the radiating sources.	225
10.1	A snapshot of the electric field near a dipole.	229
10.2	Circuit model of antenna with transmission line.	229
10.3	The distribution of current on dipole antennas.	230
10.4	Half-wave dipole represented as a series of ideal dipoles.	230
10.5	Dipole and coordinate system.	231

10.6   The folded dipole.   233
10.7   Magnetic dipole antenna and practical realisation.   234
10.8   Biconical and bow tie dipoles.   235
10.9   Gain surface of a short dipole.   237
10.10 Gain of a short dipole in a cross-section through the
       dipole axis.   238
10.11 Gain pattern of a half wavelength dipole.   239
10.12 Gain pattern of a 5/4 wavelength dipole.   239
10.13 Image current concept.   240
10.14 Connection between monopole and dipole antennas.   240
10.15 Shortening a monopole antenna.   241
10.16 Broadband conical and discone antennas.   241
10.17 Ground reflections from a dipole antenna element.   242
10.18 The effect of a finitely conducting plane.   242
10.19 Antennas located over a lossy plane.   243
10.20 Matching a short monopole to a standard load.   244
10.21 A balanced antenna when fed by (a) balanced and (b) unbalanced
       transmission lines.   245
10.22 Dipoles with coaxial feeds.   246
10.23 A half-wave dipole with offset feed.   246
10.24 A pair of phased dipoles.   247
10.25 Gain pattern of two dipoles in broadside configuration.   248
10.26 Gain pattern of two dipoles in endfire configuration.   249
10.27 Gain pattern of dipoles with 90° phasing and $\lambda/4$ spacing
       ($f = 150$ MHz).   250
10.28 The mutual impedance of half-wave dipoles in parallel and collinear
       configurations.   250
10.29 Gain pattern of five dipoles in broadside configuration.   251
10.30 A five-element array of dipoles.   251
10.31 Vertical pattern of broadside configuration.   252
10.32 Gain pattern of five dipoles in endfire configuration.   253
10.33 Gain pattern of five endfire dipoles with reduced spacing and phase
       increment.   253
10.34 Array factors for linear arrays with equally spaced elements.   255
10.35 Log periodic dipole antenna.   255
10.36 Typical gain pattern of an LPDA in a plane perpendicular to the antenna.   256
10.37 Typical gain pattern in the plane of the LPDA.   256
10.38 A general Yagi–Uda parasitic array antenna.   257
10.39 Beverage travelling wave antenna.   258
10.40 Helical antenna construction and gain.   259
10.41 Waveguide antenna and analysis.   260

10.42 A horn antenna and a reflector antenna.   262
10.43 Electric field between plates at resonance.   262
10.44 A practical patch antenna with microstrip feed.   263
11.1   The excitation of a waveguide.   268
11.2   Interaction of two antennas.   270
11.3   Communication system with two stations.   271
11.4   Propagation with ground reflections.   272
11.5   Propagation through an urban environment.   273
11.6   Propagation over a stretch of water.   274
11.7   Diffractive propagation over a building.   275
11.8   Huygen's principle and diffraction over a screen.   276
11.9   Propagation obscured by a screen.   276
11.10 The refraction of a plane wave at a plane interface.   278
11.11 The direction of propagation with refraction by a continuously varying
       medium.   278
11.12 Propagation through a spatially varying medium.   280
11.13 A ray path in a spatially varying medium.   280
11.14 Ducting propagation in the atmosphere.   281
11.15 The ionospheric F layer.   282
11.16 Ray paths in the ionosphere.   283
11.17 Ground wave propagation over a hill.   286
11.18 The scattering caused by a flat plate.   288
11.19 Total power scattered by a plate (normalised on the maximum power).   289
11.20 Propagation by scatter from a permittivity anomaly.   290
11.21 Scintillation caused by permittivity irregularity.   291
12.1   The frequency content of a real signal.   294
12.2   The sampling of an analogue signal.   294
12.3   Three signals with identical DFT due to aliasing.   296
12.4   Spectral contamination caused by aliasing.   296
12.5   A finite length discrete signal filtering system.   298
12.6   A simple one-bit analogue-to-digital converter.   299
12.7   A two-bit analogue-to-digital flash converter.   299
12.8   Digitisation of a general signal.   300
12.9   Sampled signals with quantisation noise.   301
12.10 Simple digital-to-analogue converter.   302
12.11 A direct conversion receiver with baseband analogue-to-digital
       conversion.   302
12.12 A direct digitising receiver.   303
12.13 A digital oscillator.   304
12.14 A direct digital synthesiser.   305

Preface

The following text evolved out of a series of courses on radio frequency (RF) engineering to undergraduates, postgraduates, government and industry. It was designed to meet the needs of such groups and, in particular, the needs of working engineers attempting to upgrade their skills. Thirty years ago, it appeared as if the fibre optics revolution would relegate wireless to a niche discipline, and universities accordingly downgraded their offerings in RF. In the past 10 years, however, there has been a renaissance in wireless and to a point where it is now a key technology. This has been made possible by the developments in very large-scale integration (VLSI) and CMOS technology in particular. In order to meet the manpower requirements of the wireless industry, there has been a need to upgrade the status of RF training in universities and to provide courses suitable for in-service training. The applications of wireless systems have changed greatly over the past 30 years, as has the available technology. In particular, there is a greater use of digital technologies, and antenna systems can often be of the array variety. The current text has been written with these changes in mind and there has been a culling of some traditional material that is of limited utility in the current age (graphical design methods for example). Material in the book has been carefully chosen to provide a basic training in RF and a springboard for more advanced study. At the author's own institution, RF engineering is now taught in a unified manner that emphasises the relationship between the individual components and the total system. This is part of an international trend that is gathering pace and is necessitated by the total integration of modern RF systems. The student is now expected to be able to appreciate the operation of a total system including electronics, antennas and propagation. Unfortunately, many of the textbooks covering these individual areas are extremely advanced and inappropriate at the intermediate level to which this text is mainly directed. In particular, propagation is ignored almost totally at this level. The current text is designed to service the needs of a broad RF training and to equip the reader with sufficient knowledge to appreciate the more advanced texts of other authors. As a minimum, this book requires the reader to have a basic foundation in electronics and electro-magnetism. At the author's institution the material in the first five chapters forms the basis of an initial course in RF and this is followed by a more advanced course that is based on Chapters 6 to 11.

Acknowledgements

The author would like to thank Greg Harmer, Brian Ng and Kiet To for their invaluable help with the LaTeX and diagrams for this book. He would also like to thank Said Al-Sarawi and Ken Sarkies for reading through some of the drafts. Finally, he would like to thank Alex Sharpe for his copy editing work and valuable comments.

1 Basic concepts

Broadly speaking, radio frequency (RF) technology, or wireless as it is sometimes known, is the exploitation of electromagnetic wave phenomena in that part of the spectrum between 3 Hz and 300 GHz. It is arguably one of the most important technologies in modern society. The possibility of electromagnetic waves was first postulated by James Maxwell in 1864 and their existence was verified by Heinrich Hertz in 1887. By 1895, Guglielmo Marconi had demonstrated radio as an effective communications technology. With the development of the thermionic valve at the end of the nineteenth century, radio technology developed into a mass communication and entertainment medium. The first half of the twentieth century saw developments such as radar and television, which further extended the scope of this technology. In the second half of the twentieth century, major breakthroughs came with the development of semiconductor devices and integrated circuits. These advances made possible the extremely compact and portable communications devices that resulted in the mobile communications revolution. The size of the electronics continues to fall and, as a consequence, whole new areas have opened up. In particular, spread spectrum communications at gigahertz frequencies are increasingly used to replace cabling and other systems that provide local connectivity.

The purpose of this text is to introduce the important ideas and techniques of radio technology. It is assumed that the reader has a basic grounding in electromagnetic theory and electronics. This text is not intended to be a comprehensive description of radio, but rather to provide the reader with sufficient knowledge to be able to appreciate the more advanced literature on the subject. The current chapter introduces some basic ideas concerning radio frequency systems and subsequent chapters address various aspects of the technology.

1.1 Radio waves

For a static electric field, resulting from a bounded system of charge, the field strength E will fall away at least as fast as the inverse square of the distance R from the source ($E \propto R^{-2}$). If the system suddenly changes from one static configuration to another,

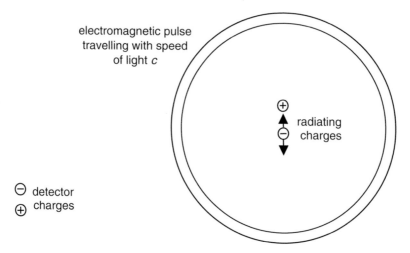

Figure 1.1   Radiation of a radial pulse.

the effect of this change will travel out from the system as a pulse. The pulse will travel with the speed of light ($c = 3 \times 10^8$ m/s) and hence, after time t, will be at distance $R = ct$ from the source. Remarkably, the amplitude of the pulse will only fall away as R^{-1} and this means that, at large distances, the effect of the pulse far outweighs that of the static fields with which it is associated. In principle, the system of charge can be *modulated* so that the fields carry away information as a series of suitably spaced pulses. As the pulses pass a second system of charge, the electric field will set these charges in motion and the pulses can be detected. The important point is that information can be carried over great distances due to the slow rate at which the pulse amplitudes decay. A dynamic electric field cannot exist in isolation and the electromagnetic theory of Maxwell predicts that there will be a magnetic field pulse associated with the electric field pulse. In fact, the electric field (E) of the pulse will be related to the magnetic field (H) through

$$E = \eta_0 H, \tag{1.1}$$

where η_0 is the impedance of the free space (377 Ω). Furthermore, the electric field, magnetic field and propagation direction will be mutually orthogonal (Figure 1.2). Far from the system of charge

$$E = \frac{K(t - R/c)}{4\pi R}, \tag{1.2}$$

where $K(t)$ is a function that depends upon the source modulation (possibly a series of pulses).

In general, any accelerating charge will give rise to a field that only falls away as R^{-1} and this leads to a great variety of ways in which fields, suitable for radio communication, can be generated. In particular, charges can be made to oscillate at a

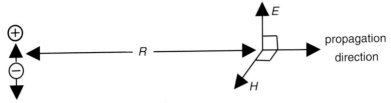

Figure 1.2   Electromagnetic field generated by a system of accelerating charge.

particular frequency ω (radians per second or $f = \omega/2\pi$ in terms of hertz) and this gives rise to a field for which function $K(t)$ is of the form $A\cos(\omega t + \phi)$ (A controls the amplitude of the field and ϕ its phase). The advantage of generating such a field is that it can allow many communications systems to coexist by operating them at different frequencies. Detection systems can be *tuned* to a particular frequency in order to receive signals from a particular source. In order to transmit information, the source will need to be *modulated* in some sense by this information. The modulating signal itself will occupy a range of frequencies known as the *baseband* and this will have a *bandwidth* that depends on the data rate. Modulation will cause the transmitted signal to spread in frequency around the purely sinusoidal *carrier* and, depending on the type of modulation, this spread can be wider than the bandwidth of the original baseband signal. Consequently, the type of modulation is an important consideration in allocating frequencies to users of the radio spectrum. Modulation is achieved by varying either A for amplitude modulation (AM), ω for frequency modulation (FM) or ϕ for phase modulation (PM).

The generation of electromagnetic waves can be illustrated through the evolution of the electric field of a simple oscillating dipole (a pair of vibrating charges with equal magnitude and opposite signs). The electric field lines will evolve as shown in Figure 1.3 (note that the field lines travel outwards from the source at the speed of light). When the charges pass, the field lines will join together and break away to make room for new field lines to develop (note that field lines cannot cross). After one period of oscillation, the field lines joining the charges will look identical to those at the start. The actual field lines, however, will have moved a *wavelength* λ out from the charges ($\lambda = c/f$ where f is the frequency of oscillation in hertz). As the oscillations continue, the field lines will continue to move out by a distance λ for each period of oscillation.

In a realistic communications system, the *radiation* fields will be produced by an electronic source that drives current into a metallic structure known as an antenna. At the atomic level, the antenna will consist of a complex combination of simple oscillating dipoles whose fields will combine to form a pattern of radiation that is dictated by the geometry of the antenna. If the metallic structure is a rod that is driven at its centre, we will have a *dipole antenna* (so called because the resulting radiation pattern is almost identical to that of a pair of oscillating charges). The field (and hence the radiated power) will be maximum in directions orthogonal to the axis of the antenna

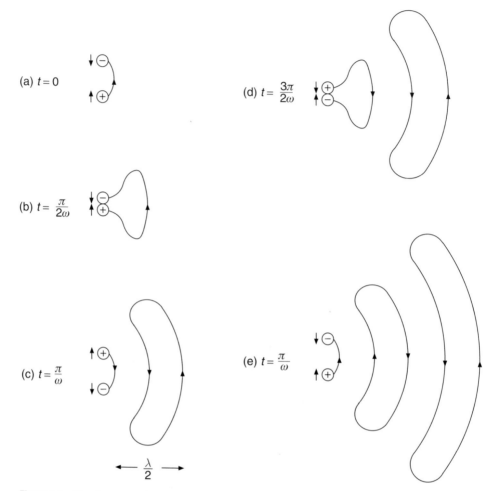

(a) $t=0$

(b) $t=\dfrac{\pi}{2\omega}$

(c) $t=\dfrac{\pi}{\omega}$

(d) $t=\dfrac{3\pi}{2\omega}$

(e) $t=\dfrac{\pi}{\omega}$

$\dfrac{\lambda}{2}$

Figure 1.3   Development of oscillating dipole field.

and minimum (zero, in fact) in directions along the axis of the dipole (note that the radiation will be symmetric about the axis). A useful way of describing the radiation characteristics of an antenna is through its *directivity* (or the related concept of *gain*). The directivity D in a particular direction is defined by

$$\text{directivity} = \frac{\text{power radiated in a particular direction}}{\text{average of power radiated in all directions}}. \tag{1.3}$$

Unfortunately, not all the power supplied to an antenna will manifest itself as radiation and some will be lost as heat in its structure (and possibly in its surroundings). The efficiency η with which the power is converted to radiation is defined by

$$\text{efficiency} = \frac{\text{total power radiated}}{\text{total power supplied}}. \tag{1.4}$$

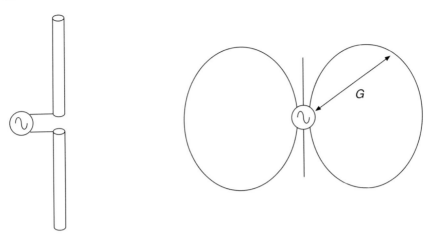

Figure 1.4   A dipole antenna and its gain pattern.

A more realistic measure of antenna effectiveness is the gain G, defined by

$$\text{gain} = 4\pi \, \frac{\text{power radiated into a unit solid angle}}{\text{total power supplied}} \qquad (1.5)$$

and, from the above definitions, it will be noted that gain $= \eta \times$ directivity. Directivity describes the deviation of radiation properties (usually expressed in dB terms) away from those of an ideal isotropic antenna and is usually represented through its *directivity pattern*. For a general antenna, the directivity pattern is a three-dimensional surface that surrounds the origin (associated with the antenna). In a particular direction, the distance of the surface from the origin is the value of directivity in that direction. The *gain pattern* of an antenna is defined in a similar fashion and is more often quoted. The directivity pattern for a half wavelength dipole is shown in Figure 1.4 (a slice through the dipole axis) and for which it should be noted that the maximum directivity is approximately 5/3 or 2.2 dBi (dB over isotropic).

From a circuit viewpoint, a transmit antenna can be represented by the circuit shown Figure 1.5. Note the $R_T + jX_T$ is the impedance of the RF source (the transmitter or Tx) and $R_r + R_L + jX_L$ is the impedance of the antenna. The antenna has two resistance contributions, R_L which represents the ohmic (heating) losses in the antenna and R_r which represents the losses due to power radiated away from the antenna (good loss). From the circuit model, it will be noted that the power P_r radiated by the antenna will be given by

$$P_r = \frac{V_T^2 R_r}{2(R_T + R_r + R_L)^2 + 2(X_T + X_A)^2}. \qquad (1.6)$$

This power will take its maximum value ($V_T^2/8R_r$) when $X_T = -X_A$ and $R_r = R_T + R_L$. A dipole is resonant (no reactance X_A in its impedance) when its length is about 0.47λ (λ is the wavelength c/f at the operating frequency f). At resonance, the radiation

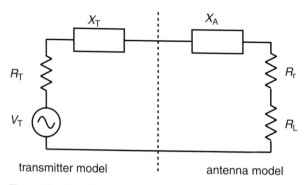

transmitter model   antenna model

Figure 1.5   Circuit model of a transmit system.

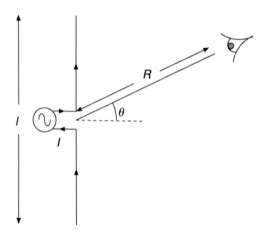

Figure 1.6   Conventions for effective length.

resistance is about 73 Ω and the loss resistance R_L is usually negligible for most practical dipoles (this means that the directivity and gain will be almost identical). Dipoles that are much shorter than a wavelength, however, have a very much smaller radiation resistance and a large capacitive reactance. For short dipoles, the radiation and ohmic resistances can often be comparable in magnitude and hence make such antennas inefficient as radiators (the directivity and gain will be significantly different in this case).

It is interesting to note that the amplitude of the electric field for any sinusoidally excited antenna has the form

$$E = \frac{\omega \mu I h_{\text{eff}}}{4\pi R} \tag{1.7}$$

at a distance R from the antenna. Quantity I is the current at the antenna feed and quantity h_{eff} has the dimensions of length and is known as the *effective length* of the antenna. For a dipole antenna, the effective length is approximately $0.64 l \cos \theta$, where l is the geometric length and θ is the angle of observation when measured from the

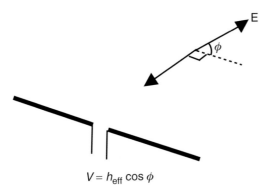

$V = h_{eff} \cos \phi$

Figure 1.7   A dipole antenna used to collect energy from an electromagnetic wave.

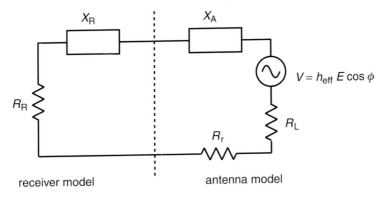

$V = h_{eff} E \cos \phi$

receiver model   antenna model

Figure 1.8   Circuit model of receive system.

plane of maximum radiation (the plane through the antenna feed that is orthogonal to the antenna axis). Now consider the case of a dipole that is used to collect energy from a time varying electric field E (the receive mode). It can be shown that an electric field will induce an open circuit voltage $h_{eff} E \cos \phi$ in the dipole terminals, where ϕ is the angle between the field direction and the axis of the antenna (the θ used in calculating h_{eff} will be the angle between the source direction and the plane that is orthogonal to the antenna axis).

The above considerations bring us to the concept of polarisation. It will be noted that the electric field can point in any direction, just so long as it is orthogonal to the direction of propagation. That is, the wave can have many different *polarisations*. A receive antenna, however, will only extract the maximum power when it is *polarisation matched* to the incoming wave ($\phi = 0$). Consequently, in a communication system, we will normally design for polarisation match. In the case of a system that uses dipole antennas for both receive and transmit, this will require the dipoles to be parallel.

A circuit model for a receiving antenna is shown in Figure 1.8. From this model, it will be noted that an antenna exhibits the same impedance $R_r + R_L + jX_A$ in both

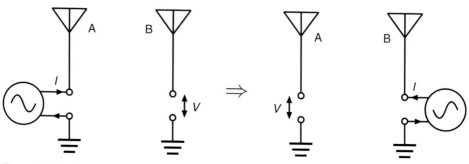

Figure 1.9   Reciprocity principle.

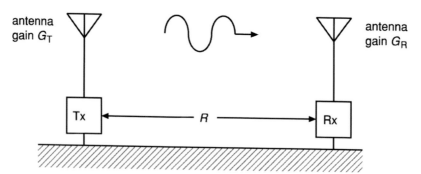

Figure 1.10   Transmit/receive system.

receive and transmit modes (maximum power will be received when $X_R = -X_A$ and $R_R = R_r + R_L$). A further result concerning the interchangeability of transmit and receive properties is known as *reciprocity*. Consider two antennas (A and B) with antenna A driven by current I and causing an open circuit voltage V in antenna B. If we now drive antenna B with the same current I and measure the open circuit voltage in A, we will find it to be the same voltage V. This is the case, even if A and B are completely different antennas. An important consequence of this result is the ability to infer two-way communication properties from one-way properties. To investigate the coverage of a mobile communications base station, for example, it is only necessary to investigate coverage of signals transmitted from the base station.

We have already noted that the field of a radio wave will fall away as the inverse of distance from the transmitter. As a consequence, the question arises as to the level of transmit power that is required in order to achieve a given level of power at the receiver (or Rx). This value can be calculated using what is known as the *Friis* equation. If we have communications between a transmitter and a receiver, distance R apart, the received power P_R and transmitted power P_T will be related through

$$P_R = P_T \left(\frac{\lambda}{4\pi R} \right)^2 G_R G_T, \tag{1.8}$$

where G_R and G_T are the gains of the receive and transmit antennas, respectively. (As can be seen from this equation, gain also measures the effectiveness of an antenna in its receive role.) The Friis equation, and its variants, are extremely important tools in the design of a radio system.

1.2   Noise

It might seem that we could transmit at any level of signal power and simply introduce a suitable amount of amplification at the receiver end. Unfortunately, this is not the case due to the fact that the signal will be competing with an ever present environment of random signals or *noise*. For example, a simple resistor will create a noise voltage v_n due to the random thermal motion of its electrons and this can be shown to have an rms voltage that satisfies

$$\overline{v_n^2} = 4kTBR, \tag{1.9}$$

where T (in kelvin) is the absolute temperature, B (in hertz) is the bandwidth of the measurement, R (in ohms) is the resistance and k is the Boltzmann constant (1.38×10^{-23} joules per kelvin). Equation 1.9 will still apply to a general impedance Z providing R is interpreted as the resistive part of the impedance (i.e., $R = \Re\{Z\}$). From a modelling viewpoint, the noise source can be regarded as an ideal voltage source of magnitude v_n in series with a noise free impedance. Alternatively, it can be regarded as an ideal current source of magnitude i_n in parallel with the impedance (note that $\overline{i_n^2} = 4kTBG$, where $G = \Re\{Z^{-1}\}$). In general, the noise in an electronic circuit can be modelled by removing the noise sources from within the circuit and replacing them by equivalent current and voltage sources at the input (Figure 1.11b). These equivalent sources can be quite complex since a general circuit can contain other forms of noise besides that due to the resistance (the shot and flicker noises of semiconductor devices, for example). In a radio receiver, the input signal will already be in competition with *external noise* from man-made sources (ignition interference, for example) and natural sources (lightning,

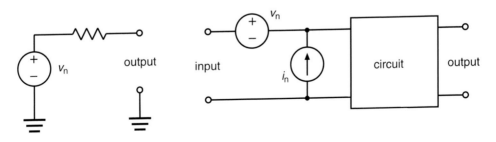

(a) Simple resistor noise source   (b) Equivalent noise sources for a complex circuit

Figure 1.11   Noise sources.

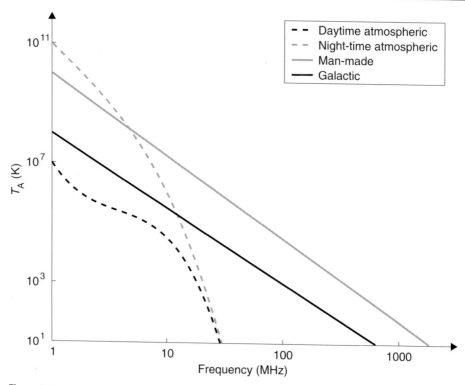

Figure 1.12   Typical antenna noise temperatures for a dipole and a variety of sources.

for example). Consequently, the input signal will need to be at a level well above that of the combined internal and external noise. It is possible to view the external noise as that arising from the antenna resistance $R_r + R_L$. To arrive at the correct level of noise, however, it is necessary to regard the system as located in an environment with an *antenna temperature* T_A that could be vastly different from the ambient temperature (around 290 K). Figure 1.12 shows some typical antenna temperatures resulting from a variety of external noise sources (note that the sources are uncorrelated and so we can simply add the temperatures to get the combined effect). In general, a noise source that is non-thermal can be treated as a thermal source with a suitably chosen *noise temperature*. External noise is the ultimate constraint since this is not under the control of the designer. For best performance, a radio receiver should be designed such that it is *externally noise limited* (i.e., the internal noise is below the expected level of external noise).

The maximum noise power that can be derived from a resistor R will be $N = kTB$ and this will be achieved when the load has impedance R. If an RF circuit is fed from a noisy source, it is clear that the amount of noise that reaches its output will depend on the circuit bandwidth B. The circuit itself will, however, add to this noise and it is important to ascertain whether the combined noise will swamp any desired signal that is present at the input. The crucial quantity in assessing circuit performance is the signal-to-noise ratio (SNR), defined by

$$\text{SNR} = \frac{S}{N} = \frac{\text{signal power}}{\text{noise power}}. \tag{1.10}$$

In a radio receiver, SNR will directly relate to the quality of the demodulated signal. The change in SNR through an RF circuit is normally measured in terms of its *noise factor F* (known as the *noise figure* when expressed in dB terms). This is defined by

$$F = \frac{S_i/N_i}{S_o/N_o}, \tag{1.11}$$

where S_i and S_o are the signal powers at the input and output, respectively, and N_i and N_o are the corresponding noise powers (limited to contributions from within the circuit bandwidth B). It can be seen that the noise factor is effectively the total noise at the output when scaled upon the noise at the output with the circuit noise sources turned off. As a consequence, the noise factor will be related to the equivalent noise sources and source impedance through

$$F = 1 + \frac{\overline{v_n^2} + \overline{i_n^2}|Z_s|^2}{4kTBR_s}, \tag{1.12}$$

where $R_s = \Re\{Z_s\}$ is the resistive part of the source impedance Z_s (note that we have assumed there to be no correlation between the voltage and current noise sources). In terms of the noise factor, the circuit noise can be approximated by a voltage source v_n at the circuit input with

$$\overline{v_n^2} = 4kTBR_s(F - 1). \tag{1.13}$$

If the source has a noise temperature T_s that is different from the ambient value T, the total noise power (referred to the input) is given by $N = kTB(F - 1) + kT_sB$ (note that noise powers are additive for uncorrelated noise sources).

For efficient communications, it is clear that the transmitted signal should be kept to the lowest level consistent with the noise environment. The signal level at the receiver antenna terminals can be calculated using the Friis equation, but this signal must compete with a combination of receiver and external noise sources. As a signal passes through the circuits of the receiver, the SNR will change through contributions from their various noise sources. In most circumstances, the received signal will be very weak and therefore will need to pass through several stages of amplification. If we cascade two stages (Figure 1.13), the noise factor of the combined device will be given by

$$F = F_1 + \frac{F_2 - 1}{G_1}, \tag{1.14}$$

where F_1 and F_2 are the noise factors of the separate stages and G_1 and G_2 are their power gains. An important deduction from this result is that the first amplifier in a

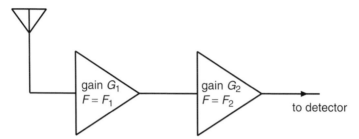

Figure 1.13   Cascaded amplifiers.

receiver needs to be the one with the best noise figure (a good first amplifier can make up for shortcomings in later stages).

Example   A 2 km radio link at 2.5 GHz uses receivers with a 10 kHz bandwidth and a noise figure of 10 dB. The system employs polarisation matched half-wave dipole antennas and there is an antenna noise temperature of 100 K. Calculate the transmitter power that is required for a 10 dB signal-to-noise ratio.

The total noise N, referred to the receiver input, is given by

$$N = (F - 1)kTB + kT_A B, \tag{1.15}$$

where $F = 10$, $T = 290\,\mathrm{K}$, $T_A = 100\,\mathrm{K}$ and $B = 10^4\,\mathrm{Hz}$. This expression includes receiver noise and antenna noise. The value of this noise power is $3.74 \times 10^{-16}\,\mathrm{W}$ and so the signal power at the receiver will need to be $P_R = 3.74 \times 10^{-15}\,\mathrm{W}$ for a 10 dB SNR. From the Friis equation, the required transmit power P_T will be

$$P_T = \frac{P_R}{G_R G_T} \left(\frac{4\pi R}{\lambda} \right)^2, \tag{1.16}$$

where $G_R = G_T = 5/3$, $R = 2 \times 10^3\,\mathrm{m}$ and $\lambda = 0.12\,\mathrm{m}$. Consequently, $P_T = 5.9 \times 10^{-5}\,\mathrm{W}$ or $-12.3\,\mathrm{dBm}$ (dB with respect to 1 mW).

The noise figure F is an extremely important parameter in the evaluation of RF device performance and its measurement is an important activity. Figure 1.14 shows one experimental set-up that can be used for this measurement. It contains a matched resistive noise source at ambient temperature T and another at an alternative temperature T_E. (The second source could be a resistor cooled to 77 K by liquid nitrogen.) For both sources, the output noise from the amplifier is measured by means of an RMS voltmeter. Let N be the noise power at the amplifier output due to the source at temperature T and N_E be that due to the source at temperature T_E, then the noise levels at the output of the device under test will be N/G and N_E/G, respectively (G is the power gain of the amplifier). As a consequence $N_E/N = [T_E + T(F - 1)]/TF$ and from which $F = (T_E/T - 1)/(N_E/N - 1)$.

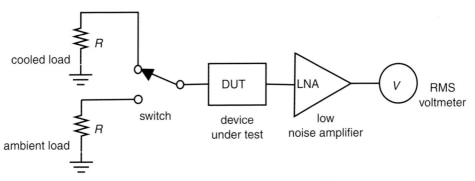

Figure 1.14   Measurement of noise figure using matched sources at different temperatures.

1.3   Sensitivity and selectivity

An important point to be noted is that the strength of the noise will depend on the bandwidth of the system. As a consequence, we need to keep the bandwidth of the receiver as low as possible (consistent with the bandwidth of the modulation). Another reason for keeping bandwidth low is the fact that all users must share a crowded spectrum and signals, other than those desired, can interfere with the detection process. A radio receiver will require some sort of filtering in order to reduce noise and discriminate against unwanted signals. The bandwidth properties, or *selectivity*, of a receiver are normally specified in terms of its 6 dB bandwidth (the bandwidth over which the response is no more than 6 dB below the maximum). Frequently, however, the 60 dB bandwidth is also included in a receiver specification.

A simple receiver system for amplitude modulated signals is shown in Figure 1.15. Output from the antenna is passed through a band-pass filter (BPF) to select the desired signal and this is then passed through a rectifier. The output of the rectifier contains a component that is proportional to the modulating signal and this is extracted by means of a low-pass filter (LPF). Such a receiver will require an extremely high quality band-pass filter if it is to operate effectively in the crowded RF environment of today. If it is required to operate over a range of frequencies, such a receiver can be impractical due to the difficulty of building high-quality variable frequency filters. Consequently, it is more usual to convert all input signals to a fixed intermediate frequency (IF). The bulk of the filtering can then take place at this IF using a filter with fixed characteristics. A receiver based on such principles is known as a superheterodyne and a typical architecture is shown in Figure 1.16 (the mixer produces a copy of the input signal that has been translated downwards in frequency by an amount equal to the oscillator frequency and this is selected by the band-pass filter). The superheterodyne is one of the many brilliant ideas pioneered by the famous radio engineer Edwin Armstrong (1890–1954).

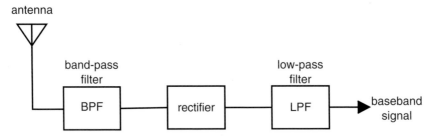

Figure 1.15   Simple receiver architecture.

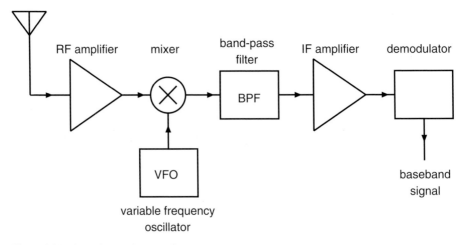

Figure 1.16   Superheterodyne receiver.

The sensitivity of a receiver is normally defined in terms of the *minimum detectable signal* (MDS), the minimum signal power that can be differentiated from the noise. This is usually taken to be the signal that yields a given signal to noise ratio S_0 (often taken to be 0 dB). For a receiver with noise factor F, and bandwidth B, the MDS is given by

$$\text{MDS} = [(F - 1)T + T_A]kBS_0, \tag{1.17}$$

where T_A is the antenna temperature and T the ambient temperature. Sometimes, the sensitivity is quoted as the signal voltage $v_{\min} = \sqrt{\text{MDS} \times R}$ that is required to achieve the signal to noise ratio S_0 (R is the input impedance of the receiver).

1.4   Non-linearity in RF systems

The introduction of amplifiers and mixers into the architecture of a receiver means that there will be non-linear effects, both intentional and unintentional, in the transfer of signals from input to the output. It is assumed that the output voltage v_o of an amplifier (or other network) is related to the input voltage v_i through

$$v_o = k_0 + k_1 v_i + k_2 v_i^2 + k_3 v_i^3 + \cdots . \tag{1.18}$$

In the case of a sinusoidal input voltage

$$v_i = V \cos \omega t \tag{1.19}$$

this will result in an output voltage of the form

$$v_o = \left(k_0 + \frac{k_2}{2}V^2\right) + \left(k_1 + \frac{3k_3}{4}V^2\right) V \cos \omega t$$

$$+ \frac{k_2}{2}V^2 \cos 2\omega t + \frac{k_3}{4}V^3 \cos 3\omega t + \cdots . \tag{1.20}$$

It will be noted that the non-linearity has generated a multitude of *harmonics*. Furthermore, the value of k_3 will normally be negative and so the gain $k_1 + (3k_3/4)V^2$ will be reduced as the input power level rises. This compression effect is normally expressed in terms of the 1 dB compression point $P_{1\,\text{dB}}$ (the input power level at which the amplifier gain is reduced by 1 dB).

If the input voltage v_i is the combination of two sinusoidal signals

$$v_i = V_1 \cos \omega_1 t + V_2 \cos \omega_2 t \tag{1.21}$$

the output will contain components at frequencies such as $\omega_1 + \omega_2, \omega_1 - \omega_2, 2\omega_1 + \omega_2,$ $\omega_1 + 2\omega_2, \ldots,$ as well as the harmonics $\omega_1, 2\omega_1, 3\omega_1, \ldots$ and $\omega_2, 2\omega_2, 3\omega_2, \ldots$ (i.e., all frequencies $m\omega_1 + n\omega_2$, where n and m are integers $0, \pm 1, \pm 2, \pm 3, \ldots$). For a non-linear behaviour with $k_n = 0$ for $n > 3$, the output voltage v_o will have the form

$$v_o(t) = k_0 + \frac{k_2}{2}\left(V_1^2 + V_2^2\right)$$

$$+ \left(k_1 V_1 + \frac{3}{4}k_3 V_1^3 + \frac{3}{2}k_3 V_1 V_2^2\right) \cos \omega_1 t$$

$$+ \left(k_1 V_2 + \frac{3}{4}k_3 V_2^3 + \frac{3}{2}k_3 V_1^2 V_2\right) \cos \omega_2 t$$

$$+ \frac{k_2}{2}V_1^2 \cos 2\omega_1 t + \frac{k_2}{2}V_2^2 \cos 2\omega_2 t$$

$$+ \frac{k_3}{4}V_1^3 \cos 3\omega_1 t + \frac{k_3}{4}V_2^3 \cos 3\omega_2 t$$

$$+ k_2 V_1 V_2 \cos(\omega_1 + \omega_2)t + k_2 V_1 V_2 \cos(\omega_1 - \omega_2)t$$

$$+ \frac{3k_3}{4}V_1^2 V_2 \cos(2\omega_1 + \omega_2)t + \frac{3k_3}{4}V_1^2 V_2 \cos(2\omega_1 - \omega_2)t$$

$$+ \frac{3k_3}{4}V_1 V_2^2 \cos(2\omega_2 + \omega_1)t + \frac{3k_3}{4}V_1 V_2^2 \cos(2\omega_2 - \omega_1)t. \tag{1.22}$$

It will be noted that the output contains components at the sum $|\omega_1 + \omega_2|$ and difference $|\omega_1 - \omega_2|$ frequencies. These components are used in the mixing process that converts an RF signal to the IF frequency in a superheterodyne receiver. Unfortunately, such a general non-linearity also produces components at frequencies that could be troublesome. If the input contains strong undesired signals at frequencies ω_1 and ω_2, the third-order intermodulation (that associated with term k_3) will produce components at frequencies $|2\omega_1 - \omega_2|$ and $|2\omega_2 - \omega_1|$. In a crowded environment, it is quite possible that some such components could be coincident with a desired frequency and hence be difficult to remove by filtering. Another effect caused by third-order terms is that of *desensitisation*. For frequency ω_1, the fundamental component at the output has amplitude $k_1 V_1 + (3/4)k_3 V_1^3 + (3/2)k_3 V_1 V_2^2$. It can be seen that, for a strong undesired signal at frequency ω_2, the response to the desired signal at frequency ω_1 will be greatly reduced when k_3 is negative. Such desensitisation can be a major problem in the presence of strong signals.

As mentioned above, the output components at frequencies $|2\omega_1 - \omega_2|$ and $|2\omega_2 - \omega_1|$ can be a major problem and it is useful to have a measure of their effect. One such measure is the intermodulation distortion (IMD) which is the ratio of the output power at the combination frequency to the output power at the fundamental. For combination frequency $|2\omega_1 - \omega_2|$, and fundamental ω_1,

$$\text{IMD} = \left(\frac{3k_3 V_1 V_2}{4k_1} \right)^2 \tag{1.23}$$

which is the same result as for combination frequency $|2\omega_2 - \omega_1|$ and fundamental ω_2. For equal input signals ($V_1 = V_2$), the input power IIP3 for which the IMD has value 1 is known as the third order intercept point at the input and is a useful measure of third-order effects (IIP3 $= |2k_1/3k_3 R|$, where R is the input impedance). A more useful practical measure of non-linearity in a receiver is the spurious free dynamic range (SFDR). This is the ratio of the minimum detectable signal to the signal whose third order distortion is just detectable. It can be shown that

$$\text{SFDR} = \left(\frac{\text{IIP3}}{\text{MDS}} \right)^{2/3}, \tag{1.24}$$

where MDS is the minimum detectable signal when measured at the receiver input. SFDR is essentially the range of signal strength for which the non-linear imperfections of the receiver will be hidden by the noise.

Example A 100 MHz receiver has a 50 Ω input impedance, a 1 dB compression point of 7 dBm (referred to the input), a bandwidth of 10 kHz and a noise figure of 6 dB. Find the spurious free dynamic range when the receiver is fed from an antenna with a noise temperature of $T_A = 1000$ K.

A 1 dB compression point of 7 dBm will imply that the non-linear voltage gain $k_1 + (3k_3/4)V^2$, when divided by the ideal voltage gain k_1, has a value $10^{-0.05}$ (1 dB of compression expressed as a voltage ratio) for an input power of 0.005 W (7 dBm expressed in watts). That is

$$\frac{k_1 + \frac{3k_3}{4}V^2}{k_1} = 10^{-0.05} \tag{1.25}$$

when $V^2/(2 \times 50) = 0.005$ W. As a consequence, $3k_3/4k_1 = -0.2175$ and so the third-order intercept point is given by IIP3 $= (1/50)|2k_1/3k_3| = 0.046$ W. The total noise at the input (including receiver noise) is given by

$$N = (F - 1)kTB + kT_A B \tag{1.26}$$

for which $F = 10^{6/10} = 4$, $T = 290$ K, $T_A = 1000$ K, $B = 10^4$ Hz and $k = 1.38 \times 10^{-23}$ J/K. As a consequence, the total noise will be $N = 2.58 \times 10^{-16}$ W which is the MDS based on a 0 dB SNR (i.e., $S_0 = 1$). The SFDR will be given by $(\text{IIP3/MDS})^{2/3} = 3.1 \times 10^9$ or 95 dB. (Note that the above analysis treats the non-linear effects as occurring before translation to baseband and assumes such a translation to be perfect.)

It is clear from the above considerations that the non-linearities of RF devices can severely degrade the performance of an RF system. As a consequence, the measurement of non-linearity is an important part of device characterisation. Because non-linearities can manifest themselves as spurious responses over a wide variety of frequencies, it is common to make measurements with a device known as a *spectrum analyser*. This is a receiver that sweeps across a range of frequencies and displays the input power as a function of frequency. Figure 1.17 shows an experimental set-up for measuring the third-order intercept point of a device. Signal sources of equal amplitude, but of different frequency (ω_1 and ω_2), are combined and then fed into the device under test (DUT). The output of the DUT is measured using a spectrum analyser to determine the output power at frequencies ω_2 and $2\omega_2 - \omega_1$. This process is then repeated for numerous input powers of increasing magnitude. The linear part of the ω_2 output curve

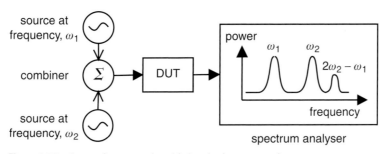

Figure 1.17   Set-up for measuring third-order intercept point.

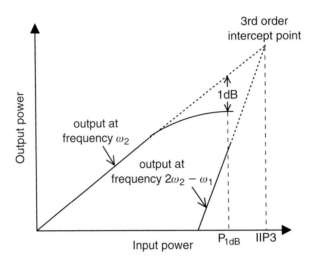

Figure 1.18   Graph depicting the third-order intercept point and 1 dB compression point.

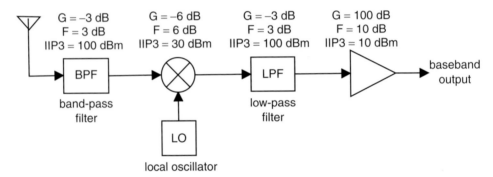

Figure 1.19   A direct conversion receiver together with the performance of its individual stages.

is then extrapolated (Figure 1.18) until it meets the $2\omega_2 - \omega_1$ output and the third-order intercept point IIP3 derived from the input power where these two lines intersect. Frequently, we will need to calculate the intercept point of a complex RF system from a knowledge of the intercept points of the constituent devices. To this end, it can be shown that the total third-order intercept point IIP3 of two cascaded devices can be derived from

$$\frac{1}{\text{IIP3}} = \frac{1}{\text{IIP3}_1} + \frac{G_1}{\text{IIP3}_2} \qquad (1.27)$$

where IIP3_1 and IIP3_2 are the intercept points for the first and second devices respectively and G_1 is the power gain of the first device.

Example   A direct conversion receiver (a receiver in which the RF signal is directly converted to baseband) is shown in Figure 1.19. The second filter of the receiver (a low-pass filter) is just wide enough to accommodate the baseband signal and the first filter is wide in order to allow the receiver to be tuned by variation of the local oscillator (LO) frequency. Using the given values of gain G, noise figure F and third order intercept point IIP3, calculate the overall gain, noise figure and third-order intercept of the system. Assuming that the final filter has a bandwidth of 3 kHz, calculate the minimum detectable signal (MDS) and dynamic range for the receiver.

Firstly, it will be noted that the noise factor F of the passive elements is given by $1/G$. This is a general result for passive two port devices and reflects the fact that components which dissipate power will also generate noise. The combined gain of the first two stages will be -9 dB and, using the expressions for combining noise factors and intercept points, the combined noise figure will be $F = 2 + (4 - 1)/(1/2) = 8$ (or 9 dB) and the combined intercept point IIP3 $= [(1/10^{10}) + (1/2)/10^3]^{-1} \approx 2000$ mW (or 33 dBm). Combining the first three stages we obtain a gain of -12 dB, a noise factor of $F = 8 + (2 - 1)/(1/8) = 16$ (or 12 dB) and an intercept point of IIP3 $= [(1/2000) + (1/8)/10^{10}]^{-1} \approx 2000$ mW (or 33 dBm). Finally, combining all stages, we obtain a total gain of 88 dB, a total noise factor of $F = 16 + (10 - 1)/(1/16) = 160$ (or 22 dB) and an intercept point of IIP3 $= [(1/2000) + (1/16)/10]^{-1} \approx 148$ mW (or 22 dBm). With MDS based on a 0 dB SNR and an ambient T_A, MDS $= FkTB = 160 \times 1.38 \times 10^{-23} \times 290 \times 3000 = 19.2 \times 10^{-16}$ W and SFDR $= [0.148/(19.2 \times 10^{-16})]^{2/3} = 1.8 \times 10^9$ or 93 dB.

1.5 Digital modulation

In digital communications, information is transferred as a stream of distinct modulation states. In binary communications, there will be two such states. These will be phase states in the case of *phase shift keying* (PSK), amplitude states in the case of *amplitude shift keying* (ASK) and frequency states in the case of *frequency shift keying* (FSK). Digital communications are highly compatible with modern computer systems, an important property considering that most communications (even voice) now take place through the intermediary of a computer system. Even before the age of the computer, and in the form of *Morse code*, digital techniques were extensively employed for the transmission of text. In Morse code, text characters are represented by combinations of spaced pulses (two different space lengths and two different pulse lengths). By adjusting the data rate, this form of communication can be made to work under conditions that would normally preclude the use of voice. Furthermore, since often used symbols are coded with smaller combinations of pulses, the transmissions are quite efficient. The advantages of Morse code are mirrored in modern digital systems in that the data rates can be adjusted to suit the noise conditions and the data coded to optimise efficiency. Another advantage of digital communications arises from the fact that a realistic communications channel will distort the transmitted signal and this can be difficult to remove in the case of analogue modulation. In the case of digital communications, however, the original transmitted signal can be reconstituted providing that the modulation states can be distinguished. In the case of binary communications, the modulating sequence $g_0 g_1 g_2 g_3 g_4 \ldots$ consists of single bit symbols (these can take the values 0 or 1) and the modulated signal (the $K(t)$ of Equation 1.2) takes the form

$$K(t) = A(t) \cos[\omega(t)t + \phi(t)], \tag{1.28}$$

where

$$A(t) = A_0 g_{[t/T]}, \quad \omega(t) = \omega_c \quad \text{and} \quad \phi(t) = 0 \tag{1.29}$$

for ASK,

$$A(t) = A_0, \quad \omega(t) = \omega_c \quad \text{and} \quad \phi(t) = \pi g_{[t/T]} \tag{1.30}$$

for PSK and

$$A(t) = A_0, \quad \omega(t) = \omega_c + \Delta f g_{[t/T]} \quad \text{and} \quad \phi(t) = 0 \tag{1.31}$$

for FSK. (Note that $[t/T]$ is the integer part of t/T, T is the duration of the pulse representing a bit, ω_c is the carrier frequency and Δf is the frequency deviation.) Whether the recovered baseband signal exhibits a 1 or a 0 will depend on whether the signal level is above or below a suitable threshold. Unfortunately, the recovered signal will always exhibit some noise and this can lead to erroneous decisions. Of all the binary modulation systems, PSK delivers the least bit error rate (10^{-5} for 10 dB of SNR) with FSK and ASK far behind (about 10^{-3} and 10^{-2}, respectively). The modulating sequence $g_0 g_1 g_2 g_3 g_4 \ldots$ can be shortened by using a quaternary base (symbols can take the values 0, 1, 2 or 3). This obviously calls for a modulation system with four states, an example of which is quadrature phase shift keying (QPSK) for which

$$A(t) = A_0, \quad \omega(t) = \omega_c \quad \text{and} \quad \phi(t) = \frac{\pi}{4}(2g_{[t/T]} + 1). \tag{1.32}$$

The important property of this modulation system is that, whilst it has the same symbol error rate as PSK, it is more efficient in that it conveys more information in a symbol.

1.6   Spread spectrum systems

The dependence of noise power upon bandwidth suggests that we should use the modulation with minimum bandwidth. There is, however, a well-known result (the Shannon–Hartley theorem) that challenges this notion. It provides a relationship between channel capacity C (bits per second), the channel bandwidth B(Hz) and the SNR. The relationship takes the form

$$C = B \log_2(1 + \text{SNR}) \tag{1.33}$$

and would appear to suggest that increasing the bandwidth will improve capacity. This, however, is moderated by the fact that the noise itself increases with bandwidth and hence the SNR decreases. Nevertheless, there are significant gains to be made by increasing bandwidth. This is the case with FM signals where it is found that the SNR of the demodulated signal increases with the frequency deviation of the modulated RF signal. Broadband FM is an example of a *spread spectrum* system, a system for which the amount of RF spectrum used is much greater than the bandwidth of the baseband signal.

Two major examples of spread spectrum systems are the frequency hopping (FH) and the direct sequence spread spectrum (DSSS) varieties. In the FH variety the system will hop around a given number of frequencies in a pseudo-random fashion that is simultaneously tracked by the receiver. The idea is that the signal will only reside in channels containing interference for a limited time and hence there will be an overall improvement in signal quality. (An added benefit is the security provided when only the intended users know the pseudo-random code.) For digital communications, such systems can be combined with error correcting codes to provide effective communications in heavily crowded environments. The alternative, a DSSS system, takes a digital baseband sequence $a(t)$ and modulates it with a much higher bit rate pseudo-random sequence $p(t)$ in order to form a new *spread* sequence $a(t)p(t)$. The spread sequence is used to modulate a carrier and hence produce an RF signal with much higher bandwidth than the original baseband signal. At the receiver end, the demodulated spread signal is multiplied by a replica of the pseudo-random sequence in order to produce the original baseband sequence. Importantly, the increased data capacity of the spread channel means that the overall power level of the transmitted signal can be significantly

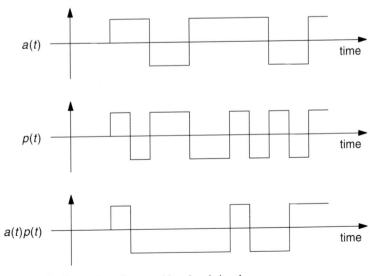

Figure 1.20   Formation of a spread baseband signal.

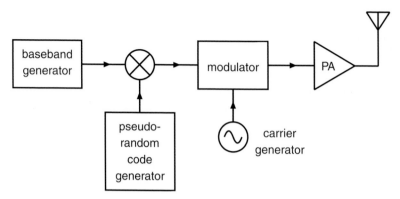

Figure 1.21   A simple DSSS transmitter.

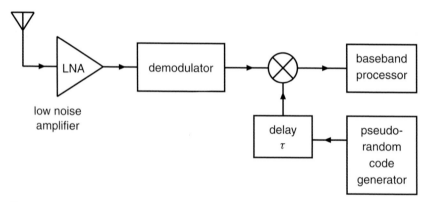

Figure 1.22   A simple DSSS receiver.

reduced. Furthermore, many users can now occupy the same channel provided they use uncorrelated pseudo-random codes (multiplication by the wrong pseudo-random code will simply produce noise).

A simple form of DSSS transmitter is shown in Figure 1.21 and a receiver in Figure 1.22. Note that there is a delay τ in the receiver in order to synchronise the pseudo-random sequences used by the receiver and transmitter. Such a delay is necessary to take account of propagation effects. Many communication channels suffer multipath propagation and this will cause there to be multiple copies of the same transmit signal, but at different delays. A DSSS system can overcome this by adding together copies of a signal that have been demodulated using several different values of delay parameter τ. If a delay does not match the signal, only noise will be produced. Successfully demodulated signals will, however, coherently add. It is important to note that this approach to the reception of DSSS signals does not require synchronisation of the receiver and transmitter. Spread spectrum techniques are increasingly used in RF systems and are to be found in the GPS navigation system and the CDMA mobile telephone system (to name but two).

1.7 Cellular radio

We have seen that spread spectrum techniques can allow several users to share a limited amount of RF spectrum. With the explosion in mobile and personal communications, however, even these techniques are insufficient to fulfil demand. The solution has been what is commonly known as *cellular radio*. In this approach, the coverage region is divided into small cells within which the transmit power is carefully limited. In this manner the same set of frequencies can be used in many different cells, providing that they are sufficiently isolated. Within each cell, the users communicate through a

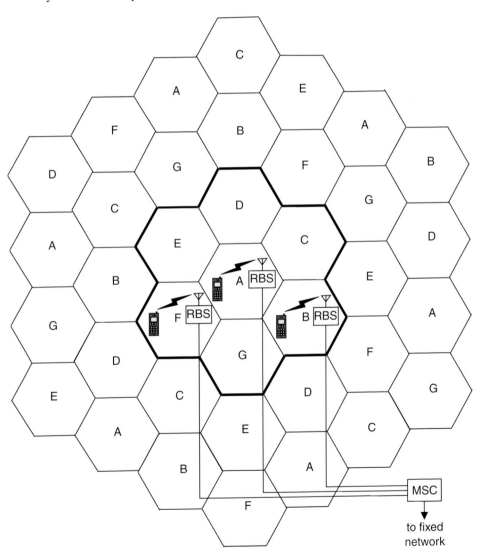

Figure 1.23 A cellular radio system.

radio base station (RBS) that is connected to other base stations, and the fixed network, through a mobile switching centre (MSC). When a user passes from one cell to another, control will be passed to the RBS associated with the new cell and the frequency will appropriately change. Modern systems are designed such that this happens in an almost seamless fashion. Cells can vary considerably in size, dependent on the propagation environment. In country areas the cells can cover tens of kilometres whilst in a city environment they might only cover one floor of a building.

A cellular system will be based around a *cluster* of cells, each of which has a unique frequency set. Figure 1.23 shows a cellular system with a cluster consisting of seven cells (the frequency sets are labelled A to G). Outside the cluster the frequency sets will be reused, but this reuse has the potential to cause interference. For a cell with frequency set A, there will be six cells just outside its cluster that reuse this frequency. (It will be noted that the minimum distance between cells with the same frequency set is approximately $4.583R$, where R is the cell radius.) Assuming that all transmitters use the same power level and that the power falls away as $(1/\text{distance})^n$, the signal to interference ratio (SIR) is given by

$$ \text{SIR} = \frac{\text{minimum power within a cell}}{6 \times \text{maximum power between cells}} = \frac{R^{-n}}{6(4.583R)^{-n}} = \frac{4.583^n}{6}. \qquad (1.34) $$

For a system in free space, n will have a value of 2. In the average cellular radio system, however, a value approaching 4 is more appropriate. As a consequence, the system of Figure 1.23 will normally provide an SIR of greater than 10 dB.

1.8 Radar systems

Radar (radio direction and ranging) forms another quite general class of RF systems that is distinct from the communications systems that we have so far discussed. Such systems transmit a pulse of RF energy whose propagation is interrupted by a *target* that causes a small amount of this energy to be returned to a receiver. The receiver will normally possess a directional antenna that allows it to ascertain the target direction. This antenna is normally a high gain reflector device that is mechanically steered towards the target or, as is increasingly the case, an array of simple antennas that is electronically steered (see Chapter 10). If the receiver and transmitter locations are known, the distance of the target can then be derived from the time of flight of the radar pulse (the pulse will travel at the speed of light c). It should be noted that the pulse length $\Delta\tau$ will dictate the bandwidth B occupied by the radar ($B \approx 1/\Delta\tau$) and also the resolution ΔR of range measurements ($\Delta R = c\Delta\tau/2$). The pulse repetition frequency (PRF) will limit the range that can be unambiguously determined since a round trip (transmitter to target and back to receiver) that takes longer than $1/\text{PRF} - \Delta\tau$ will be indistinguishable from a shorter range return from the next pulse. An alternative to

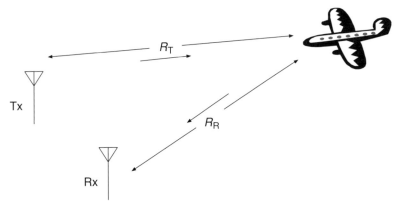

Figure 1.24   A general radar configuration.

pulsed radar is the *frequency modulated continuous wave* (FMCW) variety. In this form of radar, there is continuous transmission with the frequency swept across a bandwidth B at a suitable waveform repetition frequency (WRF). The range of the target is then ascertained from the frequency of the return. This can be achieved unambiguously to a range of $c/2$WRF and with resolution $\Delta R = c/2B$.

For radar systems, the Friis equation (known in this context as the *radar equation*) has the modified form

$$P_R = P_T \frac{\lambda^2 G_R G_T \sigma}{(4\pi)^3 R_T^2 R_R^2} \qquad (1.35)$$

where R_T and R_R represent the distances from target to transmitter and receiver respectively. Quantity σ is known as the *radar cross-section* and represents the amount of power reflected from the target when a field with unit power per unit area is incident. The cross-section is typically πa^2 for a large sphere of radius a and $0.716\lambda^2$ for a half wavelength rod. In the case that the target is moving, there will be a *Doppler shift* Δf in the frequency of the radar return. This effect can often be important in discriminating the target from unwanted returns (*clutter*) such as ground scatter. If the antennas are collocated, the connection between the Doppler shift and the component of velocity in the direction of the radar is given by

$$\Delta f = \frac{2vf}{c}. \qquad (1.36)$$

This Doppler shift is measured by observing the change in the phase of a target return over a series of pulses or sweeps (the Doppler shift is the rate of change of this phase).

Originally, radars were used to detect and locate aircraft in their vicinity. Today, however, radars are used for such diverse purposes as monitoring weather and detecting underground objects. The receiver is normally collocated with the transmitter (monostatic radar), but can sometimes be apart (bistatic radar). Broadly speaking, radar can be viewed as a radio system in which the environment provides the modulation, hence

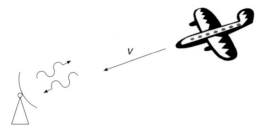

Figure 1.25 Moving target.

allowing the radar operator to infer information concerning the environment. SNR is an important issue with radar as it will limit the performance of the system. Obviously, the radar bandwidth B will dictate the level of SNR and so there is a clear trade-off between resolution and SNR. The SNR, however, can be greatly improved by the coherent addition of the returns produced by numerous successive pulses. Clutter can also severely limit the performance of a radar system and the signal-to-clutter ratio (SCR) is just as important as SNR in determining system performance.

EXERCISES

(1) Two radio stations, 5 km apart, communicate on a frequency of 300 MHz. Both stations use vertical half-wave dipole antennas with noise temperatures of 800 K. If the receivers have a 6 kHz bandwidth and a 10 dB noise figure, calculate the transmitter power that is required for 10 dB of SNR (assume an ambient temperature of 290 K).

(2) A receiver has a bandwidth of 3 kHz and a noise figure of 12 dB. Assuming a 50 Ω source at an ambient noise temperature of 290 K, calculate the sensitivity in terms of signal level (assume an MDS based on a 10 dB SNR).

(3) An amplifier has a 1 dB compression point of 10 dBm; calculate its third-order intercept point IIP3.

(4) The major noise in a receiver arises from its first mixer and preceding amplifier. If the mixer has a noise figure of 10 dB and the amplifier a noise figure of 4 dB, calculate the amplifier gain that is required to give an overall noise figure of 5 dB.

(5) A 15 MHz receiver has input impedance of 100 Ω, a 1 dB compression point of 26 dBm (referred to the input), a bandwidth of 3 kHz and a noise figure of 10 dB. Find the spurious free dynamic range (SFDR) when the receiver is fed from an antenna with noise temperature $T_A = 10^6$ K (assume an ambient temperature of 290 K).

(6) A monostatic pulse radar operates at 200 MHz with a pulse length of 0.01 s and is used to detect space junk at a distance of 300 km. The radar antennas have a gain of 12 dB and a noise temperature of 1000 K. If the debris consists of a sphere with diameter 1 m, calculate the transmit power for a radar return with 10 dB of SNR (assume the system to be externally noise limited).

SOURCES

Hayward, W. 1994. *Introduction to Radio Frequency Design*. Newark, CT: American Radio Relay League.

Krauss, J. D. 1998. *Antennas* (2nd edn). New York: McGraw-Hill.

Levanon, N. 1998. *Radar Principles*. New York: John Wiley.

Popovic, Z. and Popovic, B. D. 2000. *Introductory Electromagnetics*. Upper Saddle River, NJ: Prentice-Hall.

Pozar, D. M. 2001. *Microwave and RF Design of Wireless Systems*. New York: John Wiley.

Proakis, J. G. and Salehi, M. 1994. *Communication Systems Engineering*. Englewood Cliffs, NJ: Prentice-Hall.

Rhode, U. L., Whitaker, J. and Bucher, T. T. N. 1996. *Communication Receivers* (2nd edn). New York: McGraw-Hill.

Smith, A. A. 1998. *Radio Frequency Principles and Applications*. Piscataway, NJ: IEEE Press.

Straw, R. Dean (ed.). 1999. *The ARRL Handbook* (77th edn). Newark, CT: American Radio Relay League.

2 Frequency selective circuits and matching

Frequency selective circuits are extremely important elements of an RF system. They often consist of a combination of inductors and capacitors that achieves maximum power transfer at a particular frequency or range of frequencies. Since there will be maximum power transfer to a load when its source has conjugate impedance, such combinations will often be designed to achieve an impedance match at the frequencies to be selected. Frequency selective circuits need not be restricted to combinations of inductors and capacitors, but can also consist of lengths of transmission line or electromechanical devices such as quartz crystals. In the current chapter, we will concentrate on combinations of inductors and capacitors that have maximum transfer at a particular frequency, leaving broadband and transmission line circuits to later chapters.

Fundamental to selective circuits is the concept of *resonance*. That is, if a circuit is driven by an oscillatory stimulus, there will be a frequency, or frequencies, at which the circuit response peaks. We start the chapter by investigating this concept.

2.1 Series resonant circuits

For an inductor and capacitor in series, there will be a frequency (the resonant frequency) at which the reactance of the capacitor will cancel that of the inductor (i.e., the combination will behave as a short circuit at this frequency). Below the resonant frequency, the combination will have a capacitive reactance whose magnitude increases with decreasing frequency. Above the resonant frequency, the combination will have an inductive reactance that increases with frequency.

We will first investigate the response of a series circuit to a d.c. voltage step (see Figure 2.1). For a circuit that is switched on at time $t = 0$

$$v_i = L\frac{di}{dt} + \frac{1}{C}\int_0^t i(\tau)d\tau + Ri \tag{2.1}$$

from which, in terms of Laplace transforms (indicated by the argument s),

$$i(s) = \frac{v_i(s)}{sL + 1/sC + R}. \tag{2.2}$$

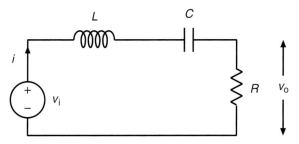

Figure 2.1   Series resonant circuit.

The voltage v_o across the load will be given by

$$v_o(s) = \frac{s(R/L)}{s^2 + Rs/L + 1/LC} v_i(s)$$
$$= A(s)v_i(s). \tag{2.3}$$

Let the circuit be switched to a constant voltage source at $t = 0$, then

$$v_i(t) = \left\{ \begin{array}{ll} 1, & t \geq 0 \\ 0, & t < 0 \end{array} \right\} = H(t) \tag{2.4}$$

for which $v_i(s) = s^{-1}$ and hence

$$v_o(s) = \frac{R/L}{s^2 + sR/L + 1/LC}$$
$$= \frac{RC}{s^2/\omega_0^2 + (2\zeta/\omega_0)s + 1}, \tag{2.5}$$

where $\omega_0 = 1/\sqrt{LC}$ is the *natural frequency* and $\zeta = (R/2)\sqrt{C/L}$ is the *damping ratio* (note that $v_o(s)$ has poles at $-\zeta\omega_0 \pm \omega_0\sqrt{\zeta^2 - 1}$). If $\zeta > 1$, the poles are real and the inverse Laplace transform yields an exponential fall in v_o. In this case, the system does not oscillate and is said to be overdamped. If $\zeta < 1$, the roots form complex conjugate pairs with a non-zero imaginary part. In this case, the inverse Laplace transform yields damped oscillatory behaviour and the system is said to be underdamped. The behaviour of $v_o(t)$ in this case will be given by

$$v_o(t) = \frac{2\zeta}{\sqrt{1 - \zeta^2}} \exp(-\zeta\omega_0 t) \sin\left(\omega_0\sqrt{1 - \zeta^2}t\right) \tag{2.6}$$

and is illustrated in Figure 2.2. Note that the circuit selects a particular frequency at which to oscillate (i.e., ring). This selective property of tuned circuits is of prime importance in RF design.

In what follows, we will often need to analyse the behaviour of a linear circuit when driven by a sinusoidal source at a particular frequency ω. Under such circumstances, we can analyse the circuit by assuming a temporal behaviour that is proportional to $\exp(j\omega t)$. For example, physical voltage can be obtained as the real part of the

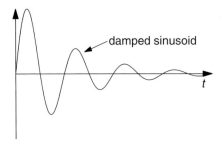

Figure 2.2   Transient response.

complex voltage

$$v(t) = V \exp(j\omega t), \tag{2.7}$$

where V is the amplitude (possibly complex). Such temporal behaviour is said to be *time harmonic* and variables with this behaviour will be denoted by lower case letters (capital letters are reserved for d.c. values and amplitudes.) It should be noted that we will sometimes denote $j\omega$ by the transform variable s, a procedure that allows us to reinterpret a time harmonic expression as a Laplace transform. In a time harmonic analysis, we can treat a capacitor C as a resistor with imaginary value $1/j\omega C$ and an inductor L as a resistor with imaginary value $j\omega L$.

In the time harmonic case, the circuit of Figure 2.1 can be regarded as a voltage divider for which

$$v_o(j\omega) = \frac{R}{j\omega L + 1/j\omega C + R} v_i(j\omega). \tag{2.8}$$

(Note that, with $j\omega$ replaced by the transform variable s, the above expression is identical to Laplace transform expression 2.3.) We define the transfer function $A(j\omega)$ of a circuit by

$$v_o(j\omega) = A(j\omega)v_i(j\omega). \tag{2.9}$$

where $v_i(j\omega)$ and $v_o(j\omega)$ are the input and output voltages, respectively. Consequently, for a series resonant circuit with time harmonic source,

$$A(j\omega) = \frac{1}{1 + jQ(\frac{\omega}{\omega_0} - \frac{\omega_0}{\omega})}, \tag{2.10}$$

where $Q = 1/\omega_0 RC = \omega_0 L/R = 1/2\zeta$. The maximum of v_o will occur at the resonant frequency ω_0 ($= 1/\sqrt{LC}$) with the response falling away either side of this frequency. Essentially, the circuit behaves as a filter with centre frequency at ω_0. The behaviour around this frequency is given by

$$A(j\omega) \approx \frac{1}{1 + \frac{2jQ}{\omega_0}(\omega - \omega_0)} \tag{2.11}$$

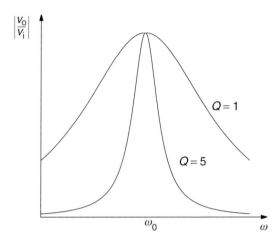

Figure 2.3 Behaviour of transfer function around resonance.

from which it is clear the value of Q controls the bandwidth (see Figure 2.3). Half power transfer occurs when $|A(j\omega)| = 1/\sqrt{2}$ and so we obtain the expression

$$\omega_{3\,\mathrm{dB}} = \omega_0 \pm \frac{\omega_0}{2Q} \tag{2.12}$$

for the 3 dB (half power) frequencies $\omega_{3\,\mathrm{dB}}$. In other words, the *quality factor* Q is related to the 3 dB bandwidth B of the circuit through $B = \omega_0/Q$. It is interesting to note that the strength of damping in the step-driven circuit is related to the bandwidth of the circuit when driven by a time harmonic source (the greater the damping the greater the bandwidth). In general, the Q of a circuit is defined by

$$Q = 2\pi \frac{\text{maximum energy stored}}{\text{energy lost per cycle}} \tag{2.13}$$

and measures the energy storage efficiency of the circuit. For the above system, the stored energy oscillates between the inductor and capacitor and so the maximum is given by $\frac{1}{2}LI_m^2$ (energy stored in an inductor L) where I_m is the maximum current in the circuit. Since the energy lost per cycle will be $\frac{1}{2}RI_m^2 2\pi/\omega$, we obtain, the expected result of $\omega L/R$ for Q.

If a source impedance is included, as shown in Figure 2.4, the bandwidth of the circuit is increased. The relationship between input and output voltages will be given by

$$v_0 = \frac{R_L}{R_L + R_S + j\omega L + 1/j\omega C} v_i = \frac{R_L}{R_L + R_S} \frac{1}{1 + jQ(\frac{\omega}{\omega_0} - \frac{\omega_0}{\omega})} v_i, \tag{2.14}$$

where the Q factor now satisfies

$$Q = \frac{\omega_0 L}{R_S + R_L}, \tag{2.15}$$

that is, the source resistance lowers Q and hence increases bandwidth.

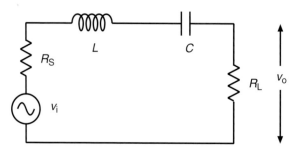

Figure 2.4 Series resonant circuit with non-ideal source.

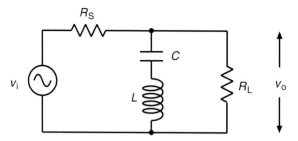

Figure 2.5 Notch filter example.

Example A $100\,\Omega$ source is connected to a $1\,\text{k}\Omega$ load with a series resonant LC circuit in parallel with the load. The LC circuit is required to remove all signals around a frequency of $150\,\text{MHz}$ and with a $3\,\text{dB}$ bandwidth of $20\,\text{MHz}$. Calculate suitable values for L and C.

The example requires us to design a *notch filter* based on a series resonant circuit (see Figure 2.5). The source and load voltages are related by

$$v_o = \frac{(j\omega L + 1/j\omega C) \parallel R_L}{R_S + (j\omega L + 1/j\omega C) \parallel R_L} v_i \tag{2.16}$$

(note that we use the notation $A \parallel B$ for the parallel combination of impedances A and B). From this expression

$$|v_o|^2 = \frac{R_L^2(1 - \omega^2 LC)^2}{(R_S + R_L)^2(1 - \omega^2 LC)^2 + R_L^2 R_S^2 \omega^2 C^2} |v_i|^2 \tag{2.17}$$

which implies a zero response to the source at a frequency of $\omega_0 = 1/\sqrt{LC}$. For small variations $\delta\omega$ around this frequency ($\omega = \omega_0 + \delta\omega$), we will have $1 - \omega^2 LC \approx -2\delta\omega/\omega_0$ and hence

$$|v_o|^2 \approx \left(\frac{R_L}{R_L + R_S}\right)^2 \frac{4\delta\omega^2}{4\delta\omega^2 + (R_L \parallel R_S)^2 \omega_0^4 C^2} |v_i|^2. \tag{2.18}$$

A 3 dB bandwidth B will require that there is half the maximum power transfer when $\delta\omega = B/2$ (maximum power transfer occurs at $\omega = 0$ for which $v_\text{o} = R_\text{L} v_\text{i}/(R_\text{L} + R_\text{S})$). For $R_\text{L} \gg R_\text{S}$ this will imply that $C = B/R_\text{S}\omega_0^2$. Noting that $\omega_\text{o} = 2\pi \times 150 \times 10^6$ rad/s and $B = 2\pi \times 20 \times 10^6$ rad/s, we will therefore need $C = 1.41$ pF and hence $L = 0.77\,\mu\text{H}$ ($L = 1/C\omega_0^2$).

2.2   Parallel resonant circuits

For a capacitor and inductor connected in parallel, there will be a frequency (the resonant frequency) at which the combination has infinite impedance (i.e., the combination behaves as an open circuit at this frequency). A circuit which illustrates this is shown in Figure 2.6.

The total impedance Z of L, C and R_L in parallel is given by

$$Z = \left(j\omega C + \frac{1}{j\omega L} + \frac{1}{R_\text{L}} \right)^{-1} = \frac{j\omega L R_\text{L}}{R_\text{L} + j\omega L - \omega^2 L C R_\text{L}} \tag{2.19}$$

so that

$$v_\text{o} = \frac{Z}{R_\text{S} + Z} v_\text{i} = \frac{j\omega L R_\text{L}}{R_\text{S} R_\text{L} + j\omega L(R_\text{S} + R_\text{L}) - \omega^2 L C R_\text{L} R_\text{S}} v_\text{i}. \tag{2.20}$$

Maximum output voltage v_o ($= R_\text{L} v_\text{i}/(R_\text{L} + R_\text{S})$) occurs at resonant frequency ω_o ($= 1/\sqrt{LC}$) for which the combined impedance of L and C is infinite. Consequently, as with the series combination of L and C, the parallel combination can be used to perform a filtering function. Around resonance,

$$v_\text{o} \approx \frac{R_\text{L}}{R_\text{S} + R_\text{L}} \frac{1}{1 + \frac{2jQ}{\omega_0}(\omega - \omega_0)} v_\text{i}, \tag{2.21}$$

where $Q = (R_\text{S} \parallel R_\text{L})/\omega_0 L$. Consequently, as for the series circuit, there will be a 3 dB bandwidth of ω_0/Q.

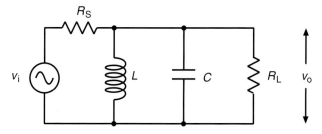

Figure 2.6   Parallel resonant circuit.

2.3 Inductive transformers

Impedance transformation is an extremely important function in that it allows us to manipulate load characteristics in order to achieve maximum power transfer. At low frequencies, this is achieved through the mutual inductance of magnetically linked inductor windings. Under ideal circumstances, the device simply transforms voltage according the ratio of the windings and current according to the inverse of this ratio. Unfortunately, at radio frequencies, the self-inductance of a transformer becomes important and the action of the transformer needs to be considered in terms of a more complex model.

For general time behaviour, the action of a two winding transformer is described by the model equations

$$v_1(t) = L_1 \frac{di_1}{dt} + M \frac{di_2}{dt}$$

$$v_2(t) = M \frac{di_1}{dt} + L_2 \frac{di_2}{dt}, \qquad (2.22)$$

where L_1 and L_2 are the self-inductances of the windings and M is their mutual inductance. For the time harmonic case, these simplify to

$$v_1(s) = sL_1 i_1(s) + sM i_2(s)$$

$$v_2(s) = sM i_1(s) + sL_2 i_2(s) \qquad (2.23)$$

and can be rearranged to yield

$$v_1(s) - \frac{M}{L_2} v_2(s) = s \left(L_1 - \frac{M^2}{L_2} \right) i_1(s)$$

$$i_1(s) + \frac{M}{L_1} i_2(s) = \frac{v_1}{sL_1}. \qquad (2.24)$$

It will be noted that, for sL_1 infinite and $M = (L_1 L_2)^{1/2}$, we obtain an ideal transformer

$$i_1 = -\frac{i_2}{n} \quad \text{and} \quad v_1 = nv_2, \qquad (2.25)$$

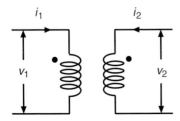

Figure 2.7   Transformer conventions.

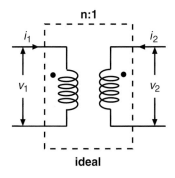

Figure 2.8   Ideal transformer.

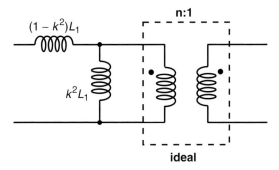

Figure 2.9   Non-ideal transformer.

where $n = (L_1/L_2)^{1/2}$ is the winding ratio (note that inductance is proportional to the square of the number of turns in the winding). Furthermore, an impedance Z_2 at terminal 2 is transformed through the transformer to an impedance $Z_1 = n^2 Z_2$ at terminal 1. (In the figures of this section, a broken box will identify an ideal transformer (Smith) and dots will identify terminals with *in phase* voltages.)

In practice, there is no such thing as an ideal transformer and the deviation from the ideal is measured in terms of the size of sL_1 and the coupling coefficient

$$k = \frac{M}{(L_1 L_2)^{\frac{1}{2}}}.$$

(2.26)

The effective turns ratio n will now satisfy

$$n = k \left(\frac{L_1}{L_2} \right)^{\frac{1}{2}}.$$

(2.27)

Noting these definitions, and Equations 2.24, it can be seen that the non-ideal transformer can be modelled as shown in Figure 2.9. If broadband operation is required, it is clear that the inductance L_1 should be as large as possible and the coupling tight ($k \approx 1$).

2.4 Tuned transformers

At radio frequencies, the self-inductance of the transformer windings can compromise the performance of a transformer. Consequently, for windings with low self-inductance, it is usual to cancel out the effect by resonating with a suitable capacitive element. Consider the configuration shown in Figure 2.10. This consists of a low impedance current source that drives a high impedance load through a transformer. In terms of an ideal transformer, this configuration is equivalent to the circuit of Figure 2.11 (assuming $k \approx 1$). For currents to balance at the load side of the transformer,

$$-\frac{i_i}{n} + v_oCs + \frac{v_o}{Ls} + \frac{v_o}{R} = 0.$$

Consequently, the mutual impedance is given by

$$Z_M = \frac{v_o}{i_i} = \frac{Ls/n}{s^2LC + sL/R + 1}$$

which has similar behaviour to that of a parallel resonant circuit with resonant frequency $\omega_0 = (LC)^{-1/2}$. At resonance, the source will see the impedance

$$Z_i = \frac{v_o/n}{i_i} = \frac{Z_M}{n} = \frac{R}{n^2}$$

and it is clear the transformer behaves in an ideal fashion.

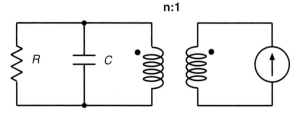

Figure 2.10   Tuned transformer.

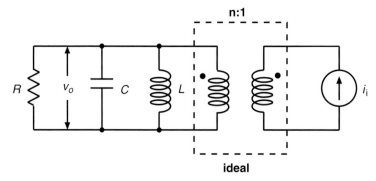

Figure 2.11   Model of tuned transformer.

2.5 Capacitive transformers

An alternative to the inductive transformer is the capacitive divider. Consider the transformation of a load through the circuit shown in Figure 2.12. The transformed impedance Z_{in} will be equal to $(1/j\omega C_1) + (1/j\omega C_2) \| R_L$ which implies the transformed admittance

$$Y_{in} = \frac{j\omega C_1 - \omega^2 R_L C_1 C_2}{j\omega R_L (C_1 + C_2) + 1}$$

$$= G_{in} + jB_{in}.$$

Assuming that load resistance R_L is much greater than the reactances of C_1 and C_2,

$$G_{in} \approx \frac{1}{R_L} \left(\frac{C_1}{C_1 + C_2}\right)^2 \quad \text{and} \quad B_{in} \approx \frac{\omega C_1 C_2}{C_1 + C_2}.$$

The addition of an inductor across the divider (as shown in Figure 2.13) will result in an input susceptance of the form

$$B_{in} \approx \frac{\omega C_1 C_2}{C_1 + C_2} - \frac{1}{\omega L}$$

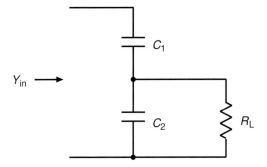

Figure 2.12 Capacitive transformer.

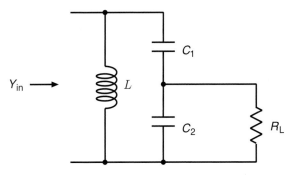

Figure 2.13 Resonant capacitive transformer.

and hence a totally real admittance Y_{in} at frequency $\omega_0^2 = (C_1 + C_2)/C_1 C_2 L$. The transformed impedance at frequency ω_0 will be $R_{in} = R_L[(C_1 + C_2)/C_1]^2$ and so the circuit will act as an ideal transformer with effective turns ratio of $n = (C_1 + C_2)/C_1$. The quality factor of the circuit is given by $Q = R_{in}/\omega_0 L$ for an ideal source (zero impedance), but this will be reduced to $Q = (R_{in} \| R_S)/\omega_0 L$ for a source with impedance R_S.

Example Design a capacitive divider to match a $100\,\Omega$ load to a $2.5\,k\Omega$ source at a frequency of $100\,MHz$ and with a bandwidth of $5\,MHz$.

Referring to Figure 2.13, $R_L = 100\,\Omega$ and $R_S = 2.5\,k\Omega$. The transformer will need an effective turns ratio of $n = \sqrt{R_S/R_L} = 5$ and so $(C_1 + C_2)/C_1 = 5$ or $C_2 = 4C_1$. The Q of the circuit will need to have a value of 20 for a $5\,MHz$ bandwidth and so $Q = (R_S \| R_{in})/\omega_0 L = 20$ which implies a value of approximately $0.1\,\mu H$ for L. To achieve a centre frequency of $100\,MHz$ we will need $(C_1 + C_2)/C_1 C_2 L = 5/4C_1 L = \omega_0^2 \approx 4 \times 10^{17}$ from which $C_1 = 31.2\,pF$ and hence $C_2 = 124.8\,pF$.

2.6   L-network matching

Although the capacitive transformer yields an effective technique for impedance transformation, there are other combinations of reactance that can perform this function. A simple means of achieving impedance transformation is through the L-network shown in Figure 2.14. Consider the problem of matching source impedance R_S to load impedance R_L. The unknown reactances X_S and X_L need to be chosen such that R_L is transformed into R_S. We will achieve this if $jX_L \| R_L$ has resistive component R_S and reactive component $-X_S$, that is

$$\frac{jX_L R_L}{R_L + jX_L} = \frac{X_L^2 R_L + jX_L R_L^2}{R_L^2 + X_L^2} = R_S - jX_S \qquad (2.28)$$

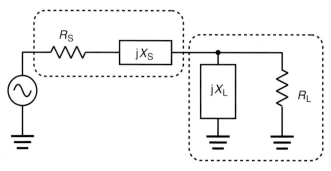

Figure 2.14   L matching network.

from which

$$R_{\text{S}} = \frac{X_{\text{L}}^2 R_{\text{L}}}{R_{\text{L}}^2 + X_{\text{L}}^2} \quad \text{and} \quad X_{\text{S}} = -\frac{X_{\text{L}} R_{\text{L}}^2}{R_{\text{L}}^2 + X_{\text{L}}^2} \tag{2.29}$$

or

$$R_{\text{S}} = \frac{R_{\text{L}}}{1 + Q^2} \quad \text{and} \quad X_{\text{S}} = -\frac{X_{\text{L}} Q^2}{1 + Q^2}, \tag{2.30}$$

where $Q = |R_{\text{L}}/X_{\text{L}}|$ is the Q of $jX_{\text{L}} \parallel R_{\text{L}}$. It will be noted that $Q = \sqrt{(R_{\text{L}}/R_{\text{S}}) - 1}$. Consequently, to match source R_{S} to load R_{L}, calculate $Q = \sqrt{(R_{\text{L}}/R_{\text{S}}) - 1}$ and then obtain the required reactances according to

$$X_{\text{L}} = \pm\frac{R_{\text{L}}}{Q} \quad \text{and} \quad X_{\text{S}} = -\frac{X_{\text{L}} Q^2}{Q^2 + 1}. \tag{2.31}$$

Note that we have assumed that $R_{\text{L}} > R_{\text{S}}$. Otherwise, we will need to reverse the L-network (i.e., design as if the source is the load and the load is the source).

The choice of whether X_{S} is capacitive or inductive depends on how the circuit is required to behave away from the *design* frequency. Reactances X_{S} and X_{L} can be viewed as forming a frequency dependent voltage divider. Since capacitor reactance will fall as frequency rises, and that of an inductor increase, the voltage output will fall if X_{L} is capacitive and rise if X_{L} is inductive. Consequently, if X_{L} is capacitive the behaviour of the L-network is that of a low-pass filter and, if inductive, the behaviour is that of a high-pass filter. Such filtering properties can be extremely useful in tailoring the frequency response of a system in which the network is employed. For example, a network is often used to match the output of a power amplifier to its load. Since (as a result of non-linearity) the amplifier can generate a significant amount of unwanted harmonics, it is often useful to configure the output network as a low-pass filter.

Thus far we have assumed a purely resistive source and load. If, however, the source and/or the load have reactive parts, the method can still be applied. We proceed as before by ignoring the reactive parts of the source and load. Once we have calculated suitable values for X_{S} and X_{L}, however, we modify these values to absorb the reactive parts of the source and load.

2.7 π- and T-networks

In some circumstances it might also be desirable for the matching network to perform the function of a band-pass filter. In such cases, the π-network would be appropriate. Figure 2.15 shows a network that is intended to match the load R_{L} to the source R_{S}. The configuration uses two L-networks to transform both source and load into a common impedance R. Because of this common impedance, the L-networks can be joined directly to form a π network. Network values can be found by solving the two

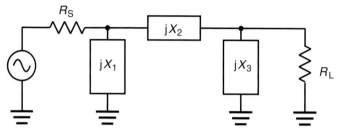

Figure 2.15 π-network.

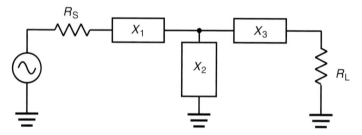

Figure 2.16   Analysis of π-network.

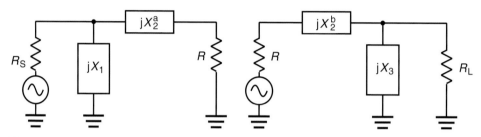

Figure 2.17   T-network.

problems illustrated in Figure 2.16. (It should be noted that we must have $R < R_S$ and $R < R_L$.)

For the calculation of bandwidth we take an overall quality factor $Q = \sqrt{[\max(R_S, R_L)/R] - 1}$, which is the highest Q of the two L-networks. R is chosen to achieve a given bandwidth (bandwidth $=$ frequency$/Q$), but it should be noted that R can be no greater than $\min(R_S, R_L)$. Both R_S and R_L are matched to R and the required value of X_2 is calculated from $X_2^a + X_2^b$.

An alternative to the π-network is the T-network. Figure 2.17 shows a T-network that is intended to match the load R_L to the source R_S. In this case, L-networks are used to transform impedances R_S and R_L into a common impedance R, but with $R > R_S$ and $R > R_L$. Because of their common impedance, the L-networks can be joined directly to form a T-network. Network reactances are found by solving the two problems illustrated in Figure 2.18.

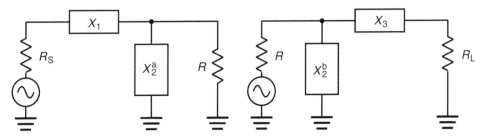

Figure 2.18   Analysis of a T-network.

For design purposes we take an overall quality factor $Q = \sqrt{[R/\min{(R_{\mathrm{S}}, R_{\mathrm{L}})}] - 1}$, which is the highest Q of the two L-networks. R is chosen to achieve a given bandwidth, but it should be noted that R must be greater than $\max(R_{\mathrm{S}}, R_{\mathrm{L}})$. Both R_{S} and R_{L} are matched to R and the required value of X_2 calculated from $X_2 = X_2^{\mathrm{a}} \parallel X_2^{\mathrm{b}}$.

2.8   Matching examples

L-network example A 100 MHz source with internal resistance $Z_{\mathrm{S}} = 25 + \mathrm{j}5\,\Omega$ is to be matched to a 50 Ω load by means of an L-network. Away from this frequency, however, the network is required to operate as a high-pass filter.

X_{S} will need to be capacitive and X_{L} inductive for a high-pass characteristic (see Figure 2.19). We will need $\mathrm{j}X_{\mathrm{L}} \parallel 50\,\Omega$ to have resistive component 25 Ω and a reactive component that cancels $5 + X_{\mathrm{S}}$. Consequently,

$$25 = \frac{X_L^2 R_L}{R_L^2 + X_L^2} \Rightarrow X_{\mathrm{L}} = 50\,\Omega \quad \text{and} \quad 5 + X_{\mathrm{S}} = -\frac{X_L R_L^2}{R_L^2 + X_L^2} \Rightarrow X_{\mathrm{S}} = -30\,\Omega,$$

where X_{L} has been chosen to be positive in order to generate a shunt inductance. Noting that $\omega = 628.3 \times 10^6$ rad/sec, the values of C and L corresponding to reactances X_{S} and X_{L} will be given by

$$C = -\frac{1}{\omega X_{\mathrm{S}}} = \frac{1}{628.3 \times 10^6 \times 30} = 53\,\mathrm{pF}$$

and

$$L = \frac{X_{\mathrm{L}}}{\omega} = \frac{50}{628.3 \times 10^6} = 79.6\,\mathrm{nH}.$$

Table 2.1 shows the performance of the filter in terms of the output power scaled upon the maximum available power (the transducer gain). It will be noted the filter has an approximately high-pass characteristic with peak power transfer at the design frequency of 100 MHz.

Table 2.1 *Performance of an L-matching network*

Frequency (MHz)	Relative power	Frequency (MHz)	Relative power
20	0.001	120	0.981
40	0.181	140	0.966
60	0.665	160	0.948
80	0.953	200	0.925
100	1.000	500	0.888

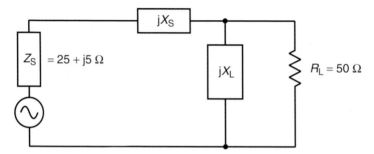

Figure 2.19   L-network example.

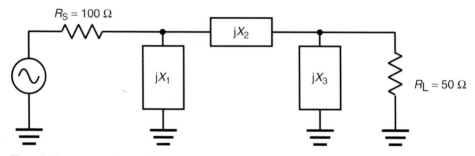

Figure 2.20 π-network example.

π-network example A 100 MHz source with 100 Ω internal resistance is to be matched to a 50 Ω load by means of a π-network. The network is to exhibit a 3 dB bandwidth of 5 MHz.

The problem is illustrated in Figure 2.20 and can be solved by converting into two auxiliary L-network matching problems (Figure 2.21). By solving these auxiliary problems, we can find the values of X_1, X_2 and X_3 (the value of X_2 is obtained from $X_2^a + X_2^b$ after solving the auxiliary problems). For a 5 MHz bandwidth at 100 MHz, we will need an overall Q of 20 (bandwidth = frequency/Q) and hence a value of

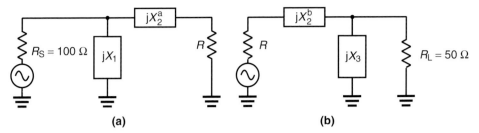

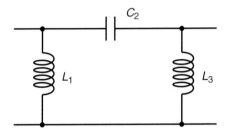

Figure 2.21   Analysis of π-network example.

Figure 2.22   Final π-network.

$0.25\,\Omega$ for R. Furthermore, we will choose X_1 and X_3 to be inductive. For network (a), $Q = \sqrt{(R_S/R) - 1} \approx 20$, so

$$X_1 = \frac{R_S}{Q} \quad \Rightarrow \quad X_1 \approx 5 \quad \text{and} \quad X_2^a = -\frac{X_1 Q^2}{Q^2 + 1} \quad \Rightarrow \quad X_2^a \approx -5.$$

For network (b), $Q = \sqrt{(R_L/R) - 1} \approx 14.1$, so

$$X_3 = \frac{R_L}{Q} \quad \Rightarrow \quad X_3 \approx 3.55 \quad \text{and} \quad X_2^b = -\frac{X_3 Q^2}{Q^2 + 1} \quad \Rightarrow \quad X_2^b \approx -3.55.$$

For the original π-network we have $X_1 = 5$, $X_2 = -8.55$ and $X_3 = 3.55$. This implies a circuit of the form shown in Figure 2.22. Since $\omega = 628.3 \times 10^6$ rad/sec, we have $L_1 = 7.95$ nH, $C_2 = 186$ pF and $L_3 = 5.65$ nH. Table 2.2 shows the performance of the filter in terms of the output power scaled upon the maximum available power. It will be noted the filter has a band-pass characteristic around the design frequency with a 3 dB bandwidth just slightly above the design value. At very high frequencies, however, the filter exhibits a high-pass characteristic.

At the end of the day, the type of matching network we choose will depend on factors such as the required frequency response and the external component reactances that need to be absorbed into the network components. Simple L-networks can only provide low- or high-pass characteristics, but combinations of them (T- and π-networks, for example) can provide band-pass filters. The principle on which the T- and π-networks were designed (the use of a common intermediate impedance) is extremely flexible and can be extended to provide matching networks with a vast range of properties.

Table 2.2 *Performance of a π-matching network*

Frequency (MHz)	Relative power	Frequency (MHz)	Relative power
70	0.002	103	0.569
90	0.046	105	0.328
95	0.140	110	0.127
97	0.413	130	0.034
98	0.608	200	0.024
99	0.841	500	0.085
100	0.993	1000	0.253
101	0.941	2000	0.543
102	0.756	5000	0.806

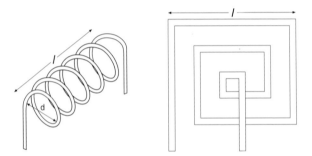

Figure 2.23   Single layer solenoid and spiral inductor.

2.9   Component reality

So far we have treated discrete components such as resistors, capacitors and inductors as having ideal behaviour. That is, in a circuit driven by a time harmonic source of frequency ω, a resistor will have a constant real resistance R (units of ohms or Ω) and present an impedance R, a capacitor will have constant real capacitance C (units of farads or F) and present an impedance $1/j\omega C$ and an inductor will have a constant real inductance L (units of henries or H) and present impedance $j\omega L$. In reality, however, all of these components will exhibit some aspects of the others in their behaviour. Arguably, the most problematic of them is the inductor.

Figure 2.23 shows two kinds of inductor, the single layer solenoid and the square spiral. For the single layer solenoid, the inductance (in μH) is given (Wheeler) by

$$L \approx \frac{d^2 N^2}{0.45d + l},$$ (2.32)

where the diameter d and length l are in units of metres and N is the number of turns. By adding a ferrite core, the inductance is multiplied by an amount equal to the relative

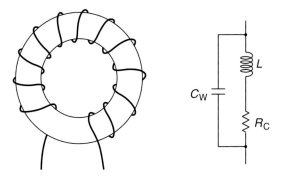

Figure 2.24   Toroidal inductor and realistic inductor model.

permeability of the core. In the case of the square spiral (see Lee),

$$L \approx 0.6N^2l, \tag{2.33}$$

where l is the length of a side.

Metal wires do not need to be shaped into a coil in order to exhibit inductance and an uncoiled length of wire (see Lee) has inductance (in μH) of

$$L = 0.2l \left[\ln \left(\frac{2l}{r} \right) - 0.75 \right], \tag{2.34}$$

where r is the radius of the wire and l is its length (an effective r of $(w + t)/\pi$ can be used for a rectangular cross-section conductor of width w and thickness t). This is an important result since nearly all components will have leads of some sort and hence exhibit some inductance in their behaviour. Figure 2.24 shows an inductor that has been wound on a toroidal core (iron dust or ferrite varieties). Such cores can greatly increase the inductance whilst introducing only limited additional loss. The inductance of the toroid can be calculated from

$$L = \frac{0.004\pi \mu A}{l} N^2 \quad \text{or} \quad L = A_L N^2. \tag{2.35}$$

In the first formula μ is the permeability of the core, A its effective cross-sectional area and l its effective length. In the second formula, the toroid properties have been collected together into a parameter A_L (normally supplied by the manufacturer). A toroidal inductor can be turned into a very effective high frequency transformer by adding an additional winding, as shown in Figure 2.25 (note that the transformer symbol now includes bars to indicate the presence of a core). An additional advantage of toroidal cores is that the field will be contained within the core and thus obviate the need for shielding. Shielding can be an important issue with inductors of the solenoid and spiral varieties. Even inductors that are quite well separated can exhibit significant

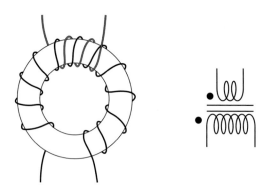

Figure 2.25   Toroidal transformer and circuit model.

mutual inductance. The inductors can be shielded, but there will be eddy currents in the shielding material and hence additional loss. Eddy currents can also be a problem for integrated circuit inductors. Such inductors will normally be of the spiral variety and sufficiently close to the substrate for there to be strong eddy currents and hence significant loss.

Besides losses due to eddy currents, there will always be resistive losses in the inductor wire. Furthermore, at high frequencies, current will tend to concentrate near the wire surface (the *skin effect*) and hence increase this loss. Yet another non-ideality will be the parasitic capacitance between the windings of the inductor. Overall, the behaviour of a real inductor can be quite complex and will need to be described by a circuit model of the sort shown in the second part of Figure 2.24. This circuit will exhibit a parallel resonance at some very high frequency and this *self-resonance* is a useful indication of the practical frequency limit of the component. Importantly, the series resistance R_C will affect the bandwidth of any circuit of which the inductor forms part. Consequently, the unloaded Q of an inductor, defined to be $Q_U = \omega L / R_C$, is a useful measure of inductor quality since it yields the maximum Q that can be attained by a circuit containing this component. Because of the large amount of eddy loss, integrated circuit inductors tend to have quite low values of Q (values around 25 are typical) and resistance R_C will often need to be taken into account. For this reason, it is useful to note that the equivalent parallel resistance is given by $Q_U \omega L$. In the example of Section 2.5, a Q of 25 for the inductor would mean a value of approximately 1.5 kΩ for the equivalent parallel resistance, a value that is almost as low as that of the parallel resistance in the ideal circuit (1.25 kΩ). As a consequence, the bandwidth of the circuit will be almost doubled.

Capacitors will usually consist of a sandwich of parallel plates separated by a dielectric (see Figure 2.26). The capacitance (units of farads) is given by $C = \epsilon A / d$, where A is the area (square metres) of the plates, d is their separation (metres) and ϵ is the permittivity of the dielectric. Parallel plates are not the only way in which capacitors can form and a pair of parallel wires can also exhibit significant capacitance. The value

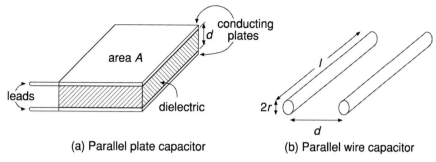

Figure 2.26   Parallel plate and parallel wire capacitors.

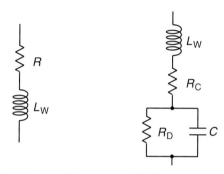

(a) Resistor model   (b) Capacitor model

Figure 2.27   High frequency circuit models of realistic resistors and capacitors.

(in farads) is given by

$$C = \frac{\pi \epsilon l}{\ln(\frac{d}{r})},$$ (2.36)

where d is the wire separation, r the wire radius and l is the length in metres. It should be noted that parallel wire capacitance is the mechanism that gives rise to the parasitic capacitance of inductors. The realistic behaviour of a capacitor is represented by the circuit model of Figure 2.27. It will be noted that the capacitor dielectric will exhibit losses that give rise to the shunt resistance R_D and ohmic losses in the leads and plates will give rise to the series resistance R_C. In addition, the leads and plates will give rise to the inductance L_w. Another component that will require a more sophisticated model at high frequencies is the resistor. A suitable model is also shown in Figure 2.27 with inductance L_w caused by the leads and body of the device.

The fact that even simple wires can form parasitic components, such as inductors and capacitors, sounds a warning for the construction of electronics at higher frequencies. Unless great care is taken in layout, the wiring will form unplanned additional components that can seriously alter the functioning of the circuit. In addition, the close proximity of inductors (or even wires) can lead to unintended mutual coupling that can cause circuit instabilities.

EXERCISES

(1) Calculate L and C such that transfer of energy is maximum at 100 MHz, and with bandwidth 10 MHz for **(a)** the circuit in Figure 2.4 with $R_S = R_L = 50 \, \Omega$ and **(b)** the circuit in Figure 2.6 with $R_S = R_L = 100 \, \Omega$.

(2) A tightly coupled transformer is required to transform a source with impedance $50 \, \Omega$ into one with an impedance of $5 \, k\Omega$ at a frequency 20 MHz. If the winding connected to the source has inductance $0.1 \, \mu H$, calculate the capacitance that needs to be placed across the secondary in order to make the circuit behave as if the transformation were perfect.

(3) It is required to connect a $50 \, \Omega$ source with frequency 100 MHz to a $5 \, k\Omega$ load by means of a capacitive transformer. Calculate the circuit values that are required for a bandwidth of 5 MHz.

(4) A 300 MHz source with impedance $10 + j5 \, \Omega$ is to be matched to a $50 \, \Omega$ load. Calculate the L-network values that will achieve this with **(a)** a low-pass characteristic and **(b)** a high-pass characteristic.

(5) A π-network is required to connect a $50 \, \Omega$ load to a 30 MHz source with impedance $200 \, \Omega$. Calculate network values that would produce a bandwidth of 3 MHz.

SOURCES

Bowick, C. 1982. *RF Circuit Design*. Boston, MA: Newnes.

Hayward, W. 1994. *Introduction to Radio Frequency Design*. Newark, CT: American Radio Relay League.

Lee, H. 1985. *The Design of RF CMOS Radio Frequency Integrated Circuits*. Cambridge University Press.

Ludwig, R. and Bretchko, P. 1990. *RF Circuit Design: Theory and Application*. Upper Saddle River, NJ: Prentice-Hall.

Smith, J. R. 1997. *Modern Communication Circuit*. New York: McGraw-Hill.

Straw, R. Dean (ed.). 1999. *The ARRL Handbook* (77th edn). Newark, CT: American Radio Relay League.

Wheeler, H. A. 1928. Simple inductive formulas for radio coils. *Proc. IRE*, **16**, 1398–1400.

3 Active devices and amplifiers

Active devices are important elements in RF systems where they perform functions such as amplification, mixing and rectification. For mixing and rectification, the non-linear properties of the device are of paramount importance. In the case of amplification, however, the non-linear properties can have a damaging effect upon performance. This chapter concentrates on small signal amplifiers for which it is possible to select conditions such that there is, effectively, linear amplification. Amplifiers based on both bipolar junction and field effect transistors are considered and the chapter includes some revision concerning their characteristics and biasing. The high frequency performance of transistor amplifiers is limited by what is known as the Miller effect and a large part of the chapter is devoted to techniques for overcoming this phenomenon.

3.1 The semiconductor diode

The semiconductor diode is a device that allows a current flow, but for which the magnitude of the flow depends in a non-linear fashion upon the applied voltage. Many varieties of diode are manufactured by forming junctions out of p- and n-type semiconductors. There are, however, several important diode varieties that have other types of junction (the semiconductor to metal junction of a Schottky diode for example). A p-type semiconductor can be formed by introducing a small amount of indium into silicon. This creates a structure that allows electrons to flow at energy levels slightly above those of the bound electrons. An n-type semiconductor can be formed by introducing a small amount of arsenic into the silicon. In this case, however, the electrons flow at energy levels much higher than those of the bound electrons. Due to the large difference in energies, it is easy to make electrons flow from n-type semiconductors into p-type, but almost impossible in the other direction.

The current i_D in a diode is related to the potential difference v_D across the diode through

$$i_D = I_s \left[\exp\left(\frac{v_D}{n V_T} \right) - 1 \right] \tag{3.1}$$

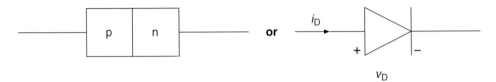

Figure 3.1 Junction diode.

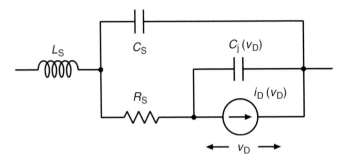

Figure 3.2 High frequency model of a diode.

when the diode is in forward bias ($v_D > 0$). The current I_s is known as the *saturation current* and is typically of the order of 10^{-15} A for a small signal diode. The *thermal voltage* V_T (typically about 25 mV) is given by $V_T = kT/q$ (where T is the absolute temperature, k is the Boltzmann constant and q is the electron charge). Parameter n is known as the *emission coefficient* and can take values between 1 and 2 (we shall assume a value of 1 in what follows as this is appropriate for diodes fabricated by integrated circuit techniques). Relation 3.1 breaks down in reverse bias ($v_D < 0$) and, under such conditions, there is very little current flow until there is sufficient reverse voltage for either *avalanche* or *Zener* breakdown.

It should be noted that a diode can exhibit significant capacitance across its junction and this can be problematic for high frequency operation. The capacitance, however, varies with the voltage across the diode and this property can be extremely useful in the construction of voltage controlled oscillators. Figure 3.2 shows a typical high frequency model of a diode. The current i_D is related to the voltage v_D through $i_D = I_s(\exp(v_D/nV_T) - 1)$ and $C_j(v_D)$ is the voltage dependent junction capacitance caused by the depletion of charge in the junction region. Capacitance C_S, inductance L_S and resistance R_S are parasitic effects caused by the diode substrate, contacts and leads.

3.2   Bipolar junction transistors

A bipolar junction transistor (BJT) is a device consisting of a sandwich of n- and p-type semiconductors (npn or pnp) that acts as a current amplifier. The device is connected through its base (B), its collector (C) and its emitter (E). It is normally constructed

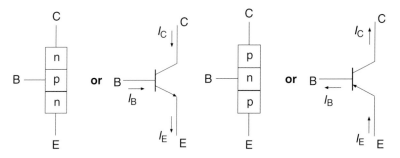

Figure 3.3 Bipolar junction transistor (BJT).

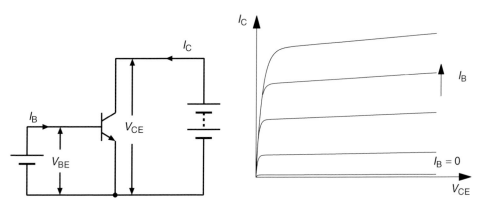

Figure 3.4 Typical BJT characteristics.

using silicon-based semiconductors, but germanium and other semiconductors can be used. For an npn device, a positive bias at the base will cause electrons to flow into this region and, if the base is thin enough, these will still have sufficient energy to surmount the base–collector junction. Figure 3.4 illustrates the characteristic performance. If the base current I_B is fixed, there is a level of emitter–collector voltage (V_{CE}) above which the relationship between collector current I_C and V_{CE} is fairly linear. This is known as the *active region*. Furthermore, in the active region, there is a fairly linear relationship between the collector and base currents for a fixed collector voltage. As a consequence, the collector and base currents are related through $I_C = \beta I_B$ where β is known as the *current gain*. The emitter–base junction will act as a diode and so the base current, and hence the collector current, will be exponentially dependent upon the emitter–base voltage V_{BE}.

Consider the situation where the transistor has a load (Figure 3.5). The current through the load will need to match the current through the transistor. Consequently, for a given base current (I_B), the collector current will be given by the intersection of the *load line* $I_C = (V_{CC} - V_{CE})/R_L$ and the relevant transistor characteristic curve. If we suitably choose the load R_L and intersection, there will be a large variation in

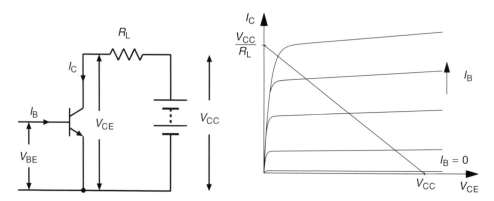

Figure 3.5 BJT load line.

collector voltage for only a small variation in base–emitter voltage (voltage amplification). We can select the correct *quiescent* state (the state around which the small variations take place) by means of suitable base *bias* conditions. For amplification, we will normally select a quiescent collector current I_C such that voltage fluctuations at the base are linearly transformed to voltage fluctuations at the load. In other applications (a mixer, for example), it is possible that we could require the transformation to be non-linear.

The most common bias configuration for small signal applications consists of a voltage divider connected to the base and a resistor in the emitter circuit (see Figure 3.6). Although the emitter resistor might seem superfluous, it plays an important role in that it provides the negative feedback that stabilises the amplifier bias against variations in transistor properties. In particular, the gain β can vary considerably between device samples and the V_{BE} required for a given collector current can vary significantly with temperature. Let I_C and I_B be the quiescent collector and base currents, respectively. The voltage drop across resistor R_2 will be $V_{BE} + (I_C + I_B)R_E$ and hence the current through this resistor will be $I_2 = [V_{BE} + (I_C + I_B)R_E]/R_2$. The current flowing through resistor R_1 will be $I_1 = I_2 + I_B$ and, noting that $V_{CC} = I_1 R_1 + I_2 R_2$, we obtain that $V_{CC} = R_1 I_B + I_2(R_1 + R_2) = R_1 I_B + (R_1 + R_2)[V_{BE} + (I_C + I_B)R_E]/R_2$. As a consequence

$$R_B V_{CC} - R_1 V_{BE} = R_1 R_B I_B + (I_C + I_B)R_1 R_E, \tag{3.2}$$

where $R_B = R_1 \parallel R_2$. Noting that $I_C = \beta I_B$, it now follows that

$$I_C = \frac{V_{CC}\frac{R_B}{R_1} - V_{BE}}{R_E + \frac{R_E + R_B}{\beta}}. \tag{3.3}$$

We will need $R_E \gg (R_E + R_B)/\beta$ for I_C to be insensitive to variations in the current gain β, that is $R_B \ll R_E(\beta - 1)$. Providing this condition is satisfied, Equation 3.3

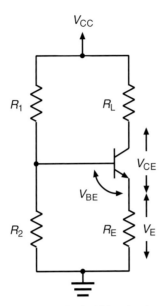

Figure 3.6   Typical bias circuit.

implies that a change of ΔI_C in I_C will require a change of ΔV_{BE} in V_{BE} where

$$\Delta I_C \approx -\frac{\Delta V_{BE}}{R_E}. \tag{3.4}$$

The base–emitter junction of a BJT behaves as a diode and so I_C will be related to the base–emitter voltage V_{BE} through $I_C \approx \beta I_s \exp(V_{BE}/V_T)$. Since V_T is temperature dependent, it can be seen that a change in temperature will cause a change in collector current I_C. From Equation 3.4, however, it is clear that a change in I_C will also cause a change in V_{BE} which counteracts the effect of the temperature change. This stabilising mechanism improves with increasing R_E, but it is normally found sufficient to choose a value of R_E that gives a V_E ($\approx I_C R_E$) of 2 V or more. Ratio $R_2/(R_1 + R_2)$ will now be fixed by $(V_{BE} + V_E)/V_{CC}$ ($V_{BE} \approx 0.7$ V for a silicon transistor). Although we need to keep R_B small in order to reduce sensitivity to variations in β, too small a value for this base resistance will end up reducing the gain of the amplifier. The usual compromise is to choose resistors R_1 and R_2 such that they carry about one tenth of the current that R_E carries. Consequently, once the quiescent collector current I_C is known, it is possible to complete the specification of R_1 and R_2. The value of the load resistance R_L needs to be chosen so that it causes a voltage drop of less than $V_{CC} - V_E$. The one-third rule, whereby V_{CE} and V_E are both approximately $V_{CC}/3$, is usually a good compromise. (If the resulting value of R_L is unacceptable, it might be necessary to relax this rule and/or readjust the bias.)

An alternative BJT bias circuit is shown in Figure 3.7, but it should be noted that the bias afforded by this configuration will not be as stable as the arrangement in Figure 3.6. If a quiescent current I_C is required, we will need a base current of

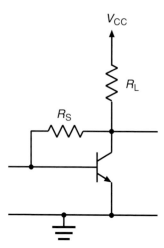

Figure 3.7 Alternative bias circuits for a BJT common-emitter amplifier.

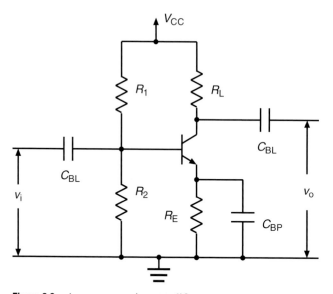

Figure 3.8 A common-emitter amplifier.

$I_B = I_C/\beta$. The total current through R_L will be $I_C + I_B = I_C(1 + 1/\beta)$ and hence $R_L = (V_{CC} - V_{CE})\beta/I_C(\beta + 1)$. In addition, we note that $R_S = (V_{CE} - V_{BE})/I_B = (V_{CE} - V_{BE})\beta/I_C$. In order to allow the maximum voltage fluctuation, we normally set $V_{CE} \approx V_{CC}/2$. Consequently, since $V_{BE} \approx 0.7$ V for silicon devices, R_L and R_S can be determined once the supply voltage V_{CC} and collector current are known.

Figure 3.8 shows a BJT amplifier in what is known as the common-emitter configuration. Blocking capacitors, labelled C_{BL}, prevent d.c. bias levels affecting the preceding or following stages and should have a small reactance (at the required operating frequency) when compared with the amplifier input and output impedances. The bypass

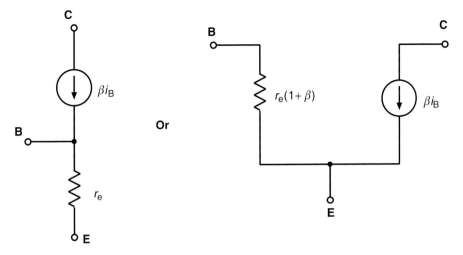

Figure 3.9 Simple BJT models.

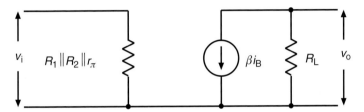

Figure 3.10 Small signal model of common-emitter amplifier.

capacitor C_{BP} short circuits the emitter resistor R_{E} at RF and should have a reactance (at the required operating frequency) that is much lower than R_{E}. It is normally assumed that the power supply V_{CC} can be regarded as grounded from the viewpoint of RF signals (zero internal impedance). This, however, is an ideal and it is normally prudent to include a suitable bypass capacitor at the V_{CC} point of the circuit. There should be at least one supply bypass capacitor per circuit section in order to avoid long and tortuous paths to ground (a possible source of harmful parasitic oscillations).

In order to analyse the behaviour of the amplifier, we will require a suitable transistor model. Since we are mainly interested in small variations around a quiescent state, linear models of the form shown in Figure 3.9 are appropriate (it should be noted that the two models are equivalent and that $r_{\mathrm{e}} \approx V_{\mathrm{T}}/I_{\mathrm{C}}$). At RF, we can replace the bypass and blocking capacitors by short circuits to obtain the small signal model of Figure 3.10 where $r_{\pi} = r_{\mathrm{e}}(\beta + 1)$. Consequently, the amplifier of Figure 3.8 will exhibit an unloaded small signal voltage gain of

$$A_{\mathrm{V}} = \frac{v_{\mathrm{o}}}{v_{\mathrm{i}}} = -\frac{R_{\mathrm{L}}\beta}{r_{\mathrm{e}}(\beta + 1)}, \tag{3.5}$$

where v_{i} and v_{o} are the RF components of the input and output voltages, respectively.

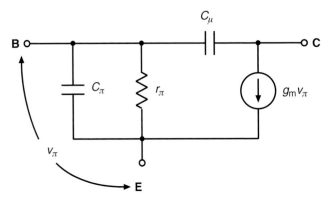

Figure 3.11 High frequency BJT model.

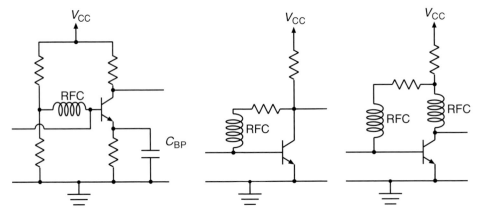

Figure 3.12 Alternative bias circuits for a BJT common-emitter amplifier.

Note that R_E without bypass would simply augment r_e and hence reduce gain. Consequently, the inclusion of C_{BP} is necessary to achieve a reasonable gain. At RF frequencies, the model of Figure 3.9 is too simplistic and needs to be modified to take account of parasitic transistor capacitances. A more appropriate model is shown in Figure 3.11 where the *transconductance* g_m can be derived from the expression $g_m r_\pi = \beta + 1$ (i.e., $g_m \approx I_C/V_T$). For most BJTs, capacitors C_π and C_μ are usually of the order of a few picofarads. The transistor capacitance will cause the current gain to fall as frequency rises and the cut-off frequency f_T (the frequency at which the short circuit common-emitter current gain of the transistor is unity) is a transistor parameter that is often quoted by manufacturers. Parameter f_T is related to C_π and C_μ through $f_T = g_m/2\pi(C_\mu + C_\pi)$.

In some circumstances, the biasing resistors can severely limit the RF performance of the circuit (by reducing Q, for example) and, because of this, the bias voltage is often delivered to the base through an RF choke (an inductor with reactance very much higher than the magnitudes of all other impedances connected to the base). The first two circuits of Figure 3.12 illustrate the use of RF chokes to isolate the bias systems described above. The third circuit of Figure 3.12 includes an additional RF choke to isolate the collector resistor. At RF frequencies, the output impedance will now be infinite and all the transistor load will come from the following stage.

3.3   The Miller effect and BJT amplifiers

The *Miller result* is extremely useful tool for understanding the constraints on transistor performance at higher frequencies. For an amplifying device, it relates a feedback impedance to equivalent input and output impedances. Consider an ideal amplifier (infinite input impedance and zero output impedance) with constant voltage gain A. For the first configuration of Figure 3.13,

$$v_i = v_o + Zi_o \tag{3.6}$$

and

$$v_o = Av_i \tag{3.7}$$

from which the effective input impedance Z_i is given by

$$Z_i = \frac{v_i}{i_o} = \frac{Z}{1 - A}. \tag{3.8}$$

In a similar fashion, the effective output impedance is given by

$$Z_o = \frac{v_o}{-i_o} = \frac{ZA}{A - 1}. \tag{3.9}$$

Consequently, the first circuit in Figure 3.13 can be replaced by the second. The additional impedance at the input is known as the *Miller impedance* and, for amplifiers with a high gain, this impedance can have a dramatic effect upon performance.

The RF behaviour of the common-emitter BJT amplifier (Figure 3.8) can be analysed through the *small signal* model of Figure 3.14 (note that $R_B = R_1 \parallel R_2 \parallel r_\pi$). Only RF voltages are considered once the quiescent conditions have been set (the blocking and bypass capacitors are replaced by short circuits). As shown in Figure 3.15, we can now

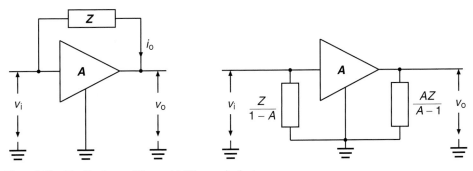

Figure 3.13   Feedback amplifier and Miller equivalent.

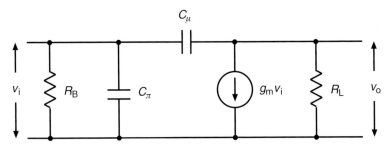

Figure 3.14   Common-emitter amplifier model.

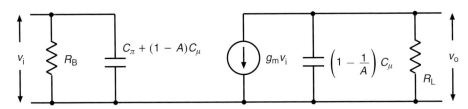

Figure 3.15   Miller transformed model.

use the Miller theorem to replace C_μ by equivalent capacitances at the input and output of the amplifier (note that $A \approx -R_L g_m$ from earlier considerations). Assuming that $|A| \gg 1$, we obtain the voltage gain

$$A_v = \frac{v_o}{v_i} = -\frac{g_m R_L}{1 + j\omega C_\mu R_L}. \tag{3.10}$$

The amplifier has the potential for high voltage gain, but this is moderated at high frequencies by the effect of C_μ.

An important consideration in design is the impedance Z_{in} that an amplifier presents to its signal source and the impedance Z_{out} that it presents to an external load (the output of the amplifier has a Thévenin equivalent circuit with impedance Z_{out} and open circuit voltage v_o). From the above model

$$Z_{in} \approx R_B \parallel \frac{1}{j\omega[C_\pi + (1 - A)C_\mu]}$$

$$= \frac{R_B}{1 + j\omega R_B[C_\pi + (1 - A)C_\mu]} \tag{3.11}$$

and

$$Z_{out} \approx \frac{1}{j\omega C_\mu} \parallel R_L. \tag{3.12}$$

In reality, the output impedance and gain will also include the effects of an additional parallel resistance r_o due to the non-ideal nature of the BJT current source ($Z_{out} \approx (1/j\omega C_\mu) \parallel R_L \parallel r_o$ and $A_v = -g_m(R_L \parallel r_o)/[1 + j\omega C_\mu(R_L \parallel r_o)]$) and this can be important in the case of a high gain amplifiers (R_L very large). A more complete transistor

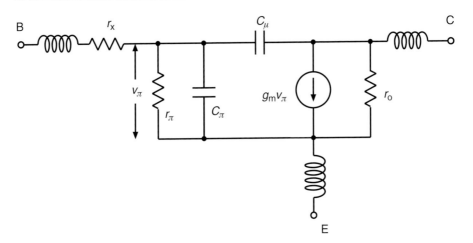

Figure 3.16 A more complete BJT model.

model that includes such effects is shown in Figure 3.16. Resistor r_o normally has a value in the order of hundreds of kilohms and the base spreading resistor r_x in the order of tens of ohms. The inductances represent the effects of transistor leads and are only significant at very high frequencies.

Concerning the input impedance Z_{in}, the major point to be noted is the strong frequency dependence caused by the *Miller capacitance* $(1 - A)C_\mu$. For a signal source with any appreciable impedance Z_S, the open circuit source voltage v_S is reduced to $v_S Z_{in}/(Z_{in} + Z_S)$ at the amplifier input and this will cause a strong fall off in amplifier performance at the higher frequencies. The additional input reactance caused by the feedback capacitance is known as the *Miller effect* and is most pronounced in high gain amplifiers. There is clearly a need for techniques to reduce its damage. For narrow band amplifiers, the simplest approach is to add an inductance that will cancel out the Miller capacitance at the required operating frequency. There are, however, amplifier configurations that can avoid the Miller effect without such a device and hence provide broadband amplification.

Example Design an npn BJT common-emitter amplifier that has a peak voltage gain of -100 for a $1\,\mathrm{k}\Omega$ collector resistor (assume $\beta = 200$, $C_\mu = 5\,\mathrm{pF}$, $C_\pi = 50\,\mathrm{pF}$ and a 9 V supply). Calculate the bandwidth of the amplifier for a $2\,\mathrm{k}\Omega$ signal source.

Referring to Figure 3.8, the voltage gain of a BJT is given by $A_v = -g_m R_L/(1 + j\omega C_\mu R_L)$ which has a peak value of $A = -g_m R_L$. Since $g_m = r_e^{-1} \approx I_C/0.026$ we will need a quiescent current of $I_C = 0.026 \times 100/R_L = 2.6\,\mathrm{mA}$ for a peak gain of -100. Noting that the voltage drop across the load resistance will be 2.6 V, the one-third rule will be approximately satisfied if we choose $R_L = R_E = 1\,\mathrm{k}\Omega$ ($I_C \approx I_E$). The base resistors (R_1 and R_2) will need to be chosen such that

$$\frac{R_1}{R_1 + R_2} = \frac{V_{BE} + V_E}{V_{CC}} = \frac{0.7 + 2.6}{9.0} = 0.367 \tag{3.13}$$

from which $R_2 = 0.58R_1$. The d.c. input resistance of the BJT will be given by $(\beta + 1)(R_E + r_e)$ which is approximately $200\,\text{k}\Omega$ and hence has negligible effect upon the base bias chain. For this chain to carry one-tenth of the current that R_E carries, we will need $V_{CC}/(R_1 + R_2) = V_E/10R_E$ and from which $R_1 + R_2 = 34.6\,\text{k}\Omega$. Consequently, $R_1 = 21.9\,\text{k}\Omega$ and $R_2 = 12.7\,\text{k}\Omega$.

Due to the small value of C_μ, the voltage gain will maintain its peak value up to quite high frequencies. The frequency $\omega_B = 1/C_\mu R_L = 2 \times 10^8\,\text{rad/s}$ (about $30\,\text{MHz}$) defines the limit of operation with a perfect voltage source (zero internal impedance). In reality, the source will have a non-zero impedance R_S and this will reduce the voltage at the amplifier input. The voltage at the input will be given by $v_i = Z_{in}v_S/(Z_{in} + R_S)$ where v_S is the open circuit source voltage and Z_{in} is the input impedance of the amplifier. Consequently, using expression 3.11 for Z_{in}, we obtain

$$v_i = \frac{R_B}{R_S + R_B}\frac{v_S}{1 + j\omega(R_S \,\|\, R_B)[C_\pi + (1 - A)C_\mu]}, \tag{3.14}$$

where $R_B = R_1 \,\|\, R_2 \,\|\, r_\pi$ (note that $r_\pi = (\beta + 1)r_e \approx 2\,\text{k}\Omega$ so that $R_B \approx 2\,\text{k}\Omega$). Since $C_\pi + (1 - A)C_\mu = 555\,\text{pF}$, it is clear that the effect of the Miller capacitance will dominate unless the source impedance is very small. Consequently, we can ignore the frequency dependence of A_v and approximate the output voltage by

$$v_o \approx \frac{R_B}{R_S + R_B}\frac{-g_m R_L v_S}{1 + j\omega(R_S \,\|\, R_B)[C_\pi + (1 - A)C_\mu]}. \tag{3.15}$$

For a transfer function of the form

$$T(s) = \frac{T_0}{1 + s/\omega_B} \tag{3.16}$$

the bandwidth is defined by ω_B (this is the frequency at which the output voltage amplitude has dropped to $1/\sqrt{2}$ of its low frequency value). Consequently, for a $2\,\text{k}\Omega$ source, the bandwidth of the amplifier can be approximated by $\omega_B = (R_S \,\|\, R_B)^{-1}[C_\pi + (1 - A)C_\mu]^{-1}$ (about $300\,\text{kHz}$). If the amplifier has an external load R_{ext}, this will reduce the low frequency amplifier gain to $A = -g_m R_L \,\|\, R_{ext}$. In turn, this will reduce the Miller capacitance and hence increase the bandwidth.

An alternative amplifier configuration, known as a common-base amplifier, is shown in Figure 3.17. Since we are interested in small signal RF performance, we can model the common-base amplifier by the circuit shown in Figure 3.18 with $R = R_3 \,\|\, r_\pi$ (we have assumed the transistor model of Figure 3.11). The voltage gain will be given by

$$A_o = \frac{v_o}{v_i} = g_m\left(R_L \,\|\, \frac{1}{j\omega C_\mu}\right) \tag{3.17}$$

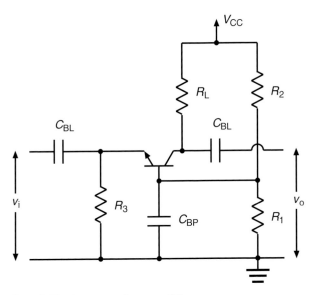

Figure 3.17   A common-base amplifier.

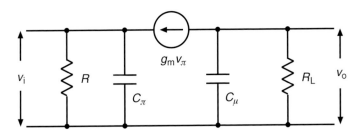

Figure 3.18   Common-base amplifier model.

and the output impedance by

$$Z_{\text{out}} = R_{\text{L}} \parallel \frac{1}{j\omega C_\mu}. \tag{3.18}$$

These quantities have a similar frequency dependence to those of the common-emitter design and likewise will require R_{L} to be replaced by $R_{\text{L}} \parallel r_{\text{o}}$ when R_{L} is large. On noting that $v_{\text{i}} = -v_\pi$, the input impedance will be given by

$$Z_{\text{in}} = \frac{\text{input voltage}}{\text{input current}} = \frac{v_{\text{i}}}{v_{\text{i}}Z^{-1} - g_{\text{m}}v_\pi} = \frac{Z}{1 + g_{\text{m}}Z}, \tag{3.19}$$

where

$$Z = R \parallel \frac{1}{j\omega C_\pi} = \frac{R}{1 + j\omega R C_\mu}.$$

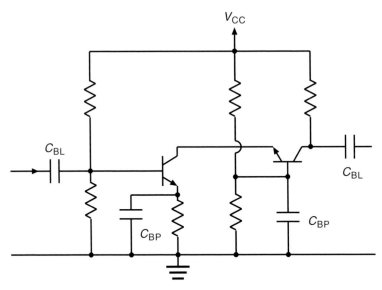

Figure 3.19　The cascode amplifier.

It is obvious that there is no Miller effect and hence a much improved frequency response compared to the common-emitter amplifier. The downside, however, is that there will be a much lower input impedance. In general, a common-base amplifier will have high voltage gain, low input impedance, high output impedance and good frequency response.

Common-emitter and common-base amplifiers can be combined to produce the cascode configuration shown in Figure 3.19. At its input this circuit has a common-emitter amplifier and hence a relatively high input impedance. This input stage, however, has a low voltage gain (the low input impedance of the common-base amplifier reduces the gain) to counteract the Miller effect. The common-base amplifier at the output of the circuit provides considerable voltage gain. As a consequence, the cascode amplifier has relatively high input and output impedances, high voltage gain and good frequency response.

Another important amplifier configuration is the emitter follower (sometimes known as a common-collector amplifier). All the previous amplifiers have provided voltage gain, but the emitter follower is different in that it only provides current gain. It has a high input impedance and low output impedance and, as a consequence, is extremely useful as a buffer for circuits (such as oscillators) that require extremely light loading. A typical configuration for the emitter follower amplifier is shown in Figure 3.20 and a suitable small signal model is shown in Figure 3.21 ($R = R_1 \parallel R_2$). It should be noted, however, that $R \parallel (1/j\omega C_\mu)$ is usually negligible in comparison with $r_\pi \parallel (1/j\omega C_\pi)$ and we will assume this to be the case in the following analysis. The currents i_1 and i_2 will satisfy

$$i_1 = (v_i - v_o)Z_\pi^{-1} \quad \text{and} \quad i_2 = \frac{v_o}{R_L}, \tag{3.20}$$

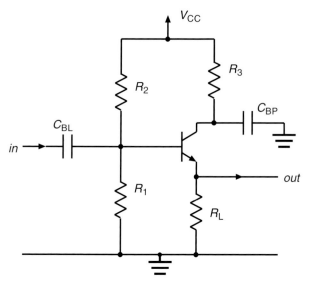

Figure 3.20   The emitter follower amplifier.

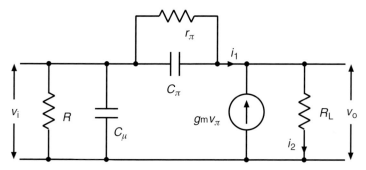

Figure 3.21   Emitter follower amplifier model.

where $Z_\pi = r_\pi \parallel (1/j\omega C_\pi)$. Since $i_1 + g_m v_\pi = i_2$, we have

$$(v_i - v_o)Z_\pi^{-1} + g_m(v_i - v_o) = \frac{v_o}{R_L} \tag{3.21}$$

from which

$$\frac{v_o}{v_i} = \frac{R_L + g_m R_L Z_\pi}{Z_\pi + R_L + g_m R_L Z_\pi} \tag{3.22}$$

and, for $|Z_\pi| g_m \gg 1$,

$$\frac{v_o}{v_i} \approx \frac{g_m R_L}{1 + g_m R_L} < 1. \tag{3.23}$$

The output impedance Z_{out} can be derived from the open circuit output voltage divided by the short circuit current. Assuming $|Z_\pi| g_m \gg 1$, the open circuit voltage will be given by $g_m R_L v_i/(1 + g_m R_L)$ and, in the same limit, the short circuit current by $g_m v_i$.

As a consequence, $Z_{\text{out}} \approx R_L/(1 + g_m R_L)$ and this will be quite a low value under normal circumstances. Caution, however, should be exercised in the case that the amplifier has a high impedance source since this can cause substantial deviations from the above value for output impedance. If Z_S is the internal resistance of the source, it is found that $Z_{\text{out}} \approx R_L \parallel [1/g_m + Z_S/(\beta + 1)]$.

The input impedance Z_{in} of the amplifier can be derived from the input voltage divided by the input current. This will yield

$$Z_{\text{in}} = \frac{v_i}{i_i} = \frac{v_i Z_\pi}{v_i - v_o} = \frac{Z_\pi}{1 - \frac{v_o}{v_i}}$$
$$= Z_\pi(1 + g_m R_L) + R_L$$

which simplifies to

$$Z_{\text{in}} = \frac{(R_L + r_e)(\beta + 1)}{1 + j\omega C_\pi r_\pi} + R_L. \tag{3.24}$$

For frequencies below $1/C_\pi r_\pi$, the value of Z_{in} is usually quite high due to the gain factor β. Although the amplifier provides no voltage gain, it is clear that it acts as an effective high to low impedance transformer with good high frequency response (no Miller effect).

3.4   Differential amplifiers

Differential amplifiers have the ability to amplify voltage differences and are the basis of many integrated circuit designs. The gain of these amplifiers can be electronically controlled and, because of this, they can also be configured as mixers. Figure 3.22 shows a typical BJT differential amplifier. If we assume the simple small signal BJT model of Figure 3.9, we obtain the following expressions for the emitter currents

$$i_a = \left[v_i^a - (i_a + i_b)R_e\right]g_m$$
$$i_b = \left[v_i^b - (i_a + i_b)R_e\right]g_m$$

which imply that

$$i_a - i_b = g_m\left(v_i^a - v_i^b\right)$$

and

$$i_a + i_b = g_m \frac{v_i^a + v_i^b}{1 + 2g_m R_e}.$$

From the above expressions, assuming a high current gain, we obtain the differential output voltage

$$v_d = v_o^a - v_o^b \approx -g_m R_c\left(v_i^a - v_i^b\right) \tag{3.25}$$

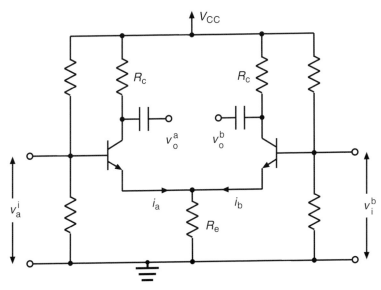

Figure 3.22 A BJT differential amplifier.

and the common output voltage

$$v_c = \frac{v_o^a + v_o^b}{2} \approx -\frac{g_m R_c}{1 + 2g_m R_e}\left(\frac{v_i^a + v_i^b}{2}\right)$$

$$\approx -\frac{R_c}{2R_e}\left(\frac{v_i^a + v_i^b}{2}\right). \tag{3.26}$$

For $R_c/R_e \ll 1$, we have a differential output alone ($v_c \approx 0$) with differential gain $-g_m R_c$. Assuming the bias resistors to have negligible effect, the amplifier will have a differential input impedance of $2r_\pi$ (r_π for each side) and a differential output impedance of $2R_c \parallel r_o$ ($R_c \parallel r_o$ for each side).

In the case of integrated circuit (IC) realisations, it is important to keep passive components down to a minimum (d.c. instead of capacitive coupling and the minimum of resistors). Passive components are difficult to fabricate on silicon and take up large amounts of real estate with a wide parameter spread. IC implementations tend to use current sources for biasing and active loads instead of resistive loads. In terms of current sources, the above differential amplifier design can be replaced by the circuit shown in Figure 3.23. The base voltage will need to have a d.c. bias that is sufficient to allow the current source to operate, but the exact value is not critical. As a consequence, suitable bias can be achieved through a single resistor to the d.c. supply.

For an ideal current source, the value of R_e will be infinite so that the common mode output will be zero. Figure 3.24 shows an example of a practical current source. The current in the left transistor is set by its base–emitter voltage ($V_{BE} = V_{CC} - I_o R$). Given the sensitivity of I_o to changes in V_{BE}, the current will be very stable when the collector resistor R is large. Since both transistors have identical base–emitter voltages,

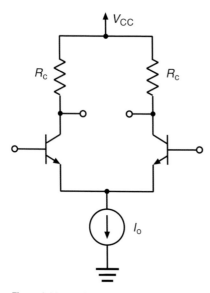

Figure 3.23 Differential amplifier with current source.

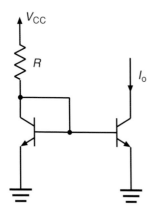

Figure 3.24 A current mirror ($I_0 = (V_{CC} - V_{BE})/R$ and $V_{BE} \approx 0.7\,V$).

the current in the right–hand transistor will be the same as that in left-hand transistor. As a consequence, the configuration is known as a *current mirror*.

In the case of IC designs, it is common to replace the collector resistor by an *active load* (some examples of which are shown in Figure 3.25). By connecting the base of a BJT to its collector (the first circuit of Figure 3.25), we can form a resistor with value equal to the emitter resistance r_e (a fairly low value). This can be developed into a high impedance load by use of the current mirror concept, as shown in the second circuit of Figure 3.25. The left-hand BJT will behave as a load of impedance r_e, but the right-hand BJT will behave as a load of impedance r_o. A differential amplifier with single-ended output and active loads is shown in Figure 3.26. For this amplifier, the output voltage will be given by $v_o = g_m v_d (R_L \parallel 2r_o)$ where R_L is the impedance of the external load. Provided R_L is large, the gain of the amplifier will be large.

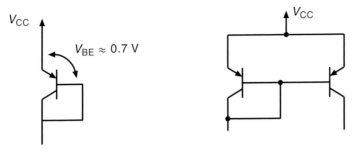

Figure 3.25   Active loads.

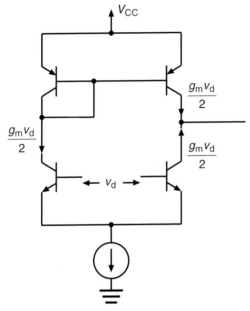

Figure 3.26   Differential amplifier with active loads (only a.c. currents shown).

It will be noted that a differential amplifier can be regarded as a pair of back-to-back common-emitter amplifiers with inputs $v_d/2$ and $-v_d/2$, respectively. As a consequence, it will suffer from the Miller effect. It can, however, be used to produce a single-ended amplifier that is immune from the Miller effect. Figure 3.27 shows a suitable circuit. The RF grounding of the Q_1 collector eliminates the Miller effect on transistor Q_1 and the RF grounding of the base of Q_2 eliminates the Miller effect on Q_2.

3.5   Feedback

We have already seen how negative feedback can be used to stabilise a BJT against variations in device characteristics. Feedback is also an important mechanism for RF circuits in that it provides a means of tailoring amplifier performance. Consider the

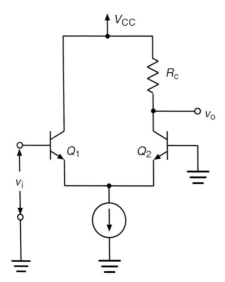

Figure 3.27   A high frequency amplifier.

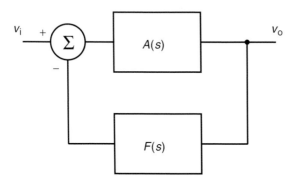

Figure 3.28   A feedback system.

negative feedback loop shown in Figure 3.28. The amplifier has a transfer function
$A(s)$ and the feedback loop has a transfer function $F(s)$ (note that the feedback is
subtracted from the input). For the above loop

$$\frac{v_o}{v_i} = A_F(s) = \frac{A(s)}{1 + A(s)F(s)} \quad (3.27)$$

from which it can be seen that feedback can be used to manipulate the transfer characteristics of an amplifier. In particular, feedback can be used to reduce the effect of the
distortion caused by amplifier non-linearity. In the limit $|AF| \gg 1$,

$$\frac{v_o}{v_i} \approx \frac{1}{F(s)}, \quad (3.28)$$

that is, the behaviour of the system is primarily dictated by its feedback characteristics.
If the feedback is a linear passive network, it is clear that a highly linear amplifier

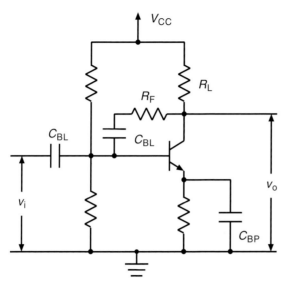

Figure 3.29 BJT amplifier with feedback.

could result (even if the basic amplifier itself is highly non-linear). Figure 3.29 shows a simple example. We assume a high gain BJT and $R_F \gg R_S$ where R_S is the combined impedance of the signal source, the bias network and the transistor. The feedback loop will have a gain of $-R_S/R_F$ and so the amplifier will have a voltage gain of approximately $-R_F/R_S$.

   Another use of feedback is to increase the bandwidth of an amplifier. Let the dominant frequency dependence of an amplifier be described by

$$A(s) = \frac{A_0}{1 + s/\omega_B} \tag{3.29}$$

where ω_B sets the bandwidth of the amplifier. Consider a feedback loop with constant gain $F(s) = A_f$, then

$$A_F(s) = \frac{A(s)}{1 + A(s)A_f}$$

$$= \left(\frac{A_0}{1 + A_0 A_f} \right) \left[\frac{1}{1 + \frac{s}{\omega_B(1+A_0 A_f)}} \right]. \tag{3.30}$$

Note that the bandwidth has increased from ω_B to $\omega_F = \omega_B(1 + A_0 A_f)$, but that the d.c. gain has decreased from A_0 to $A_F(0) = A_0/(1 + A_0 A_f)$. Interestingly, the gain–bandwidth product remains constant (i.e., $A_0\omega_B = A_F(0)\omega_F$). Because of its constancy, the gain–bandwidth product is often used as a parameter in amplifier specification. Since bandwidth is such an important consideration in RF amplifiers, it is useful to have a means of estimating its value. For a simple RC circuit, the frequency response will be dictated by the time it takes to discharge the capacitor and this leads to an estimate of $\omega_B = 1/RC$ for the bandwidth. For a more complex system, a general estimate (Sedra

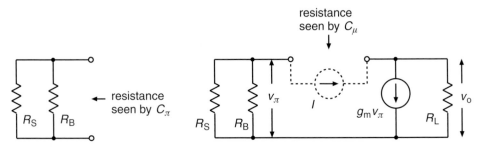

Figure 3.30   The calculation of resistances for bandwidth calculations.

and Smith) of the bandwidth is given by $\omega_B = 1/(\sum_{i=1}^{N} R_i C_i)$ where C_1 to C_N are the N capacitors of the system and R_i is the resistance that capacitor C_i will see when all other capacitors have been removed. (Note that, before such an analysis, all coupling and bypass capacitors should be replaced by short circuits and source voltages set to zero.)

Example   For the common-emitter amplifier of the previous example, use the above approximate technique to estimate the bandwidth when the source has a $2\,k\Omega$ impedance.

Referring to Figure 3.14, the resistance that capacitor C_π will see is that of the first circuit of Figure 3.30. Consequently, C_π will see the resistance of $R_\pi = R_S \parallel R_B$ (this is approximately $1\,k\Omega$ since $R_B \approx 2\,k\Omega$ from the first example). The resistance R_μ seen by C_μ is that of the second circuit. This can be found by forcing a current I to flow through the circuit. The voltage v_π across the input resistors will be $v_\pi = -I(R_S \parallel R_B)$ and, since a current $I - g_m v_\pi$ will flow through R_L, the voltage v_o across this resistor will be given by $v_o = (I - g_m v_\pi)R_L = IR_L[1 + g_m(R_S \parallel R_B)]$. Consequently, since the voltage across the source of current I will be $v_o - v_\pi$, the resistance R_μ will be given by $R_\mu = (v_o - v_\pi)/I = R_L + (1 + g_m R_L)(R_S \parallel R_B)$. Using values from the first example, we find that $R_\mu \approx 102\,k\Omega$. The bandwidth can be approximated by $\omega_B = 1/(R_\mu C_\mu + R_\pi C_\pi)$ which gives a value of 284 kHz (note that $C_\mu = 5\,pF$ and $C_\pi = 50\,pF$).

The major problem with feedback is stability. If the phase of $A(s)F(s)$ is equal to $180°$ at a frequency where $|A(s)F(s)| = 1$, then the amplifier will be unstable since an input, however small (the amplifier noise, for example), will produce an infinite output. In general, the amplifier will be unstable if $|A(s)F(s)| > 1$ when the phase of $A(s)F(s)$ is equal to $180°$. We need the phase to remain less than $180°$ up to the frequency for which $|A(s)F(s)|$ as fallen to 1. The difference between $180°$ and the phase when $|A(s)F(s)| = 1$ is known as the *phase margin* and is a useful measure of amplifier stability. It has been previously noted that the frequency response of a circuit is essentially controlled by the size and distribution of capacitance. In particular, base to collector capacitance can have a large impact through the Miller effect. By the addition of suitable capacitance within an amplifier circuit, it is often possible to modify the frequency response of the amplifier so that it remains stable in a feedback application. Such a process is known as *compensation* and is an extremely important tool in feedback amplifier design.

Although intentional feedback can be a useful tool in the design of RF circuits, unintentional feedback can be a problem. For example, the feedback afforded by the collector to base capacitance of a BJT, together with an inductive load, such as an RF choke, can often be the cause of low frequency instability (Teale). Such instabilities can give rise to *parasitic* oscillations that degrade the performance of an amplifier. It is important to ensure that an amplifier is stable on all frequencies. In particular, many low frequency parasitic oscillations can be traced to poor decoupling of the d.c. supply and this area should always be given special attention.

3.6   Field-effect transistors

A field-effect transistor (FET) is a semiconductor device in which the current flow (between source and drain terminals) is controlled by the voltage on a high input impedance terminal (the gate). The voltage at the gate causes a field that controls the width of the conduction channel from source to drain. There are a large variety of FET devices and three examples are shown in Figures 3.31–3.33.

The junction-field-effect transistor (JFET) has all its terminals connected to the semiconductors (both n- and p-type) by means of ohmic contacts. In the case of a metal oxide semiconductor FET (MOSFET), however, the gate is insulated from the semiconductor. Characteristic FET behaviour is shown in Figure 3.34. In the triode region, the current in the drain–source channel can be approximated by

$$I_D = K\left[2(V_{GS} - V_t)V_{DS} - V_{DS}^2\right] \tag{3.31}$$

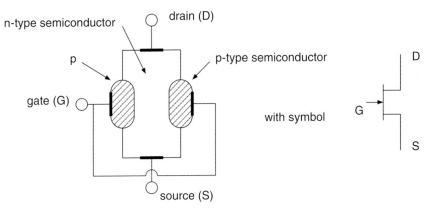

Figure 3.31   The n-channel junction field-effect transistor (JFET).

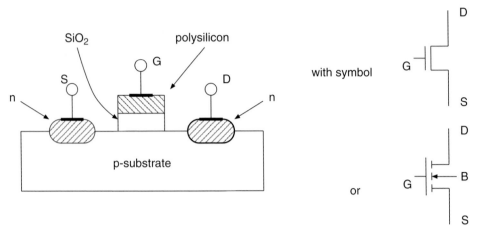

Figure 3.32   The n-channel enhancement MOSFET (nMOS).

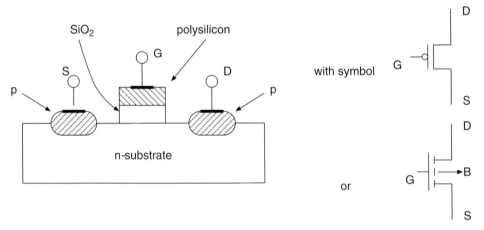

Figure 3.33   The p-channel enhancement MOSFET (pMOS).

and in the saturation region by

$$I_D = K(V_{GS} - V_t)^2, \qquad (3.32)$$

where K is a constant that depends on the device construction, V_t is a parameter known as the threshold voltage, V_{GS} is the gate–source voltage and V_{DS} is the drain–source voltage. In the above FET examples, the following should be noted:

(a) For the JFET, V_t is negative and the device requires $V_{GS} > V_t$ for it to turn on. It requires $V_{DS} > V_{GS} - V_t$ to operate in the saturation region.

(b) For the nMOS FET the device requires $V_{GS} > V_t$ for it to turn on and $V_{DS} > V_{GS} - V_t$ to operate in the saturation region (V_t will be positive in an enhancement mode device).

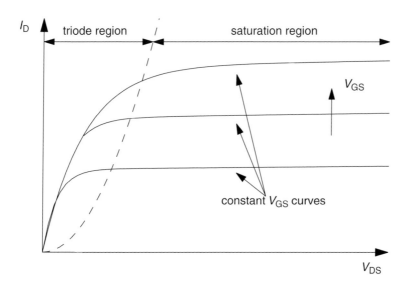

Figure 3.34   Typical FET characteristics.

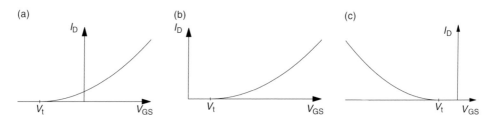

Figure 3.35   Typical $I_D - V_{GS}$ curves for (a) a JFET, (b) an nMOS enhancement FET and (c) a pMOS enhancement FET.

(c) For the pMOS FET the device requires $V_{GS} < V_t$ for it to turn on and $V_{DS} < V_{GS} - V_t$ to operate in the saturation region (V_t will be negative in an enhancement mode device).

Both nMOS and JFET devices require the drain to be positive with respect to the source, but a pMOS device requires the drain to be negative. For the above devices, the relationship between gate voltage and drain current is illustrated in Figure 3.35.

Expression 3.32 normally provides an effective description of saturated FET behaviour when the drain–source channel is long. When this channel is short, however, the current will also depend on the drain voltage. Under these circumstances, the characteristic is better modelled by

$$I_D = K(V_{GS} - V_t)^2(1 + \lambda V_{DS}), \tag{3.33}$$

where λ is known as the channel length modulation parameter.

The main problem with the FET is the large spread in device characteristics as a result of the manufacturing process. Consider the effect of this spread upon the typical

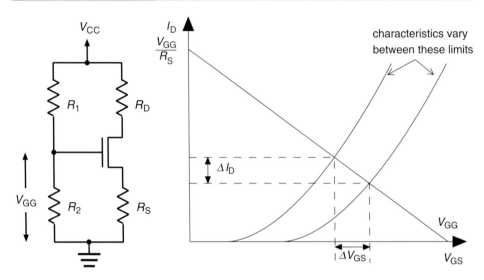

Figure 3.36   Bias configuration and bias spread characteristics.

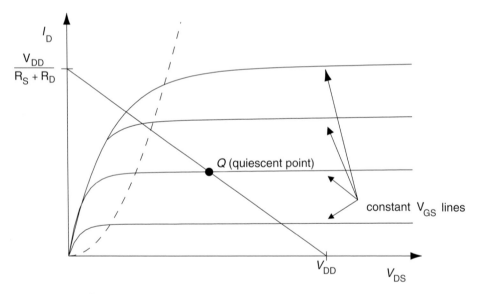

Figure 3.37   FET load line.

bias circuit (shown in Figure 3.36). If ΔI_D is an acceptable spread in drain current, the source resistor R_S should be chosen such that $R_S > \Delta V_{GS}/\Delta I_D$ where ΔV_{GS} is the spread in gate–source voltage (derived from manufacturers data) corresponding to ΔI_D. A typical FET characteristic with its load line is illustrated in Figure 3.37. For amplifier applications, the bias point will need to be well away from the triode region and it is normal to choose a combination of resistors and current I_{DQ} such that $I_{DQ}(R_S + R_D) \approx V_{DD}/2$, where V_{DD} is the supply voltage. The exact choice of R_D and I_{DQ}

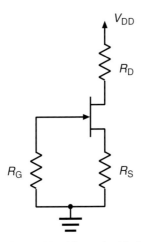

Figure 3.38 Alternative biasing for a JFET.

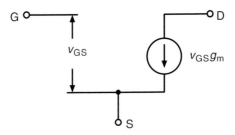

Figure 3.39 Low frequency FET model.

will depend upon the requirements, such as gain and output impedance. Once the quiescent drain current I_{DQ} has been decided upon, the corresponding gate voltage V_{GG} can be derived from the characteristic equation

$$I_{DQ} = K(V_{GG} - I_{DQ}R_S - V_t)^2. \tag{3.34}$$

Voltage V_{GG} is generated by the divider resistors R_1 and R_2 with $V_{GG} = R_2V_{DD}/(R_1 + R_2)$. The resistors should be as large as possible, but substantially less than the FET input resistance (thousands of megohms). In the case of an n-channel JFET, there is the option of self-bias (as shown in Figure 3.38). The gate is set at ground voltage through a high value resistance R_G (significantly less than the FET input resistance). R_S is chosen to give a quiescent current around which there is suffi-cient linearity for the expected input swing, but should also be large enough to guard against device variations.

Once the quiescent conditions have been set, we are normally interested in small fluctuations about this state. Under such circumstances, and providing the frequency is low, the variations can be studied using the linear model of Figure 3.39 where the transconductance g_m is derived from

$$g_m = \frac{\partial I_D}{\partial V_{GS}} \tag{3.35}$$

with the derivative evaluated at the quiescent operating point. In terms of quiescent current I_{DQ}, $g_m = 2\sqrt{K I_{DQ}}$.

3.7   FET amplifiers

A common-source amplifier is shown in Figure 3.40. In order to analyse this circuit, we will need a small signal FET model that is suitable for RF frequencies. Such a model is shown in Figure 3.41. Apart from the absence of an input resistor (its value is far too high to be considered) and the drain-to-source capacitance (usually quite small in comparison with other device capacitances), the model is identical to that for a BJT. Without the effect of capacitances, the gain will be $A_0 = -g_m(R_D \| r_d)$. Consequently, assuming $|A_0| \gg 1$, we can use the Miller theorem to transform C_{GD} into equivalent input and output capacitances. Then, by an almost identical analysis to that for the common-emitter amplifier,

$$\frac{v_o}{v_i} = \frac{-g_m(R_D \| r_d)}{1 + j\omega(R_D \| r_d)(C_{GD} + C_{DS})}. \tag{3.36}$$

Furthermore, the input impedance will take the form

$$Z_{in} = \frac{1}{j\omega[C_{GS} + (1 - A_0)C_{GD}]} \tag{3.37}$$

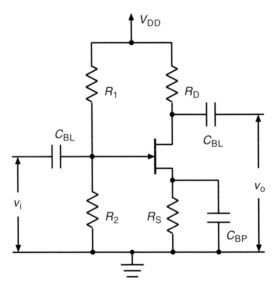

Figure 3.40   Common-source amplifier.

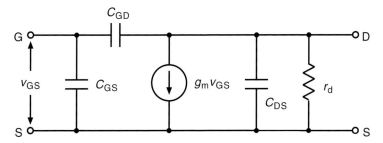

Figure 3.41   Model of an FET device at RF frequencies.

and output impedance the form

$$Z_{\text{out}} = R_D \parallel r_d \parallel \frac{1}{j\omega(C_{\text{GD}} + C_{\text{DS}})}. \tag{3.38}$$

It is clear that the Miller effect will also be a significant factor in FET amplifiers at high frequencies. There are, however, additional effects that can influence FET performance as the frequency rises. In particular, the gate-to-drain and gate-to-source capacitances of the device will include a small amount of series resistance. This will result in an input resistance that falls in value as frequency rises (typically to a value of a few kilohms at 100 MHz), an effect that must be taken into account when matching FETs at very high frequencies.

Example Design an JFET common-source amplifier that has a peak RF voltage gain of -10 and harmonic components 26 dB below the fundamental when there is a 100 mV amplitude RF drive. For the chosen transistor, and a 1 kΩ signal source, calculate the bandwidth of the amplifier.

We will design the amplifier around the 2N3819 JFET for which typical parameters are $K = 1.3 \times 10^{-3} \, \text{A/V}^2$, $V_t = -3 \, \text{V}$, $C_{\text{GS}} = 4 \, \text{pF}$, $C_{\text{GD}} = 1.6 \, \text{pF}$, $C_{\text{DS}} \approx 0$ and $r_d = 30 \, \text{k}\Omega$. The circuit is shown in Figure 3.40 and, if we choose the self-bias option, we can remove resistor R_1. Resistor R_2 will need to have a high value in order to maintain a high amplifier input impedance, but the value must be well below the d.c. input impedance of the amplifier (several megohms) in order to maintain the gate at zero d.c. A value of 100 kΩ is usually sufficient. The output voltage at the drain will have the form

$$v_o = V_{\text{DD}} - R_D K (V_{\text{bias}} + v_{\text{RF}} - V_t)^2, \tag{3.39}$$

where V_{bias} is the gate–source voltage under quiescent conditions and v_{RF} is the RF input voltage. For $v_{\text{RF}} = V_{\text{RF}} \cos \omega t$,

$$v_o = V_{\text{DD}} - R_D K (V_{\text{bias}} - V_t)^2 - 2 R_D K (V_{\text{bias}} - V_t) V_{\text{RF}} \cos(\omega t) - R_D K V_{\text{RF}}^2 \cos^2(\omega t) \tag{3.40}$$

and so the output voltage v_o will have amplitude $2R_D K(V_{bias} - V_t)V_{RF}$ at the fundamental frequency ω and $\frac{1}{2}R_D K V_{RF}^2$ at the harmonic frequency 2ω (noting that $2\cos^2(\omega t) = 1 + \cos(2\omega t)$). Consequently, we require

$$\frac{V_{RF}}{4(V_{bias} - V_t)} < \frac{1}{20} \tag{3.41}$$

for the second harmonic to be at the requisite level (26 dB implies that the harmonic voltage be $1/20$ of the fundamental voltage). This will imply that $V_{bias} > -2.5\,V$ for input signal at the level $V_{RF} = 100\,mV$. Assuming the transistor to be saturated, we obtain from the characteristic equation that

$$-V_{bias} \doteq R_S K (V_{bias} - V_t)^2. \tag{3.42}$$

We choose a bias voltage $V_{bias} = -1.5\,V$ and from Equation 3.42 obtain that $R_S = 513\,\Omega$. This should be sufficient to guard against device variation. If not, there is clearly scope to further reduce the bias without compromising the requirement on harmonic performance. On noting that $g_m = 2K(V_{bias} - V_t)$, the amplifier will have a voltage gain given by $-2R_D K(V_{bias} - V_t)$ and to attain a value of -10 will require R_D to have a value of $2.6\,k\Omega$. (Note that r_d, and any external load, will need to be incorporated into R_D.) Thus far we have neglected the effect of device capacitance, but this needs to be considered as it will set the high frequency limit of amplifier operation. As with the common-emitter amplifier, the dominant effect will come from the Miller capacitance at the input. This will combine with the finite impedance R of the signal source to reduce the amplifier input voltage. The voltage will be given by $v_i = Z_{in}v_S/(Z_{in} + R)$, where v_S is the open circuit voltage of the source and Z_{in} is the input impedance of the amplifier. From the expression 3.37 for Z_{in}, we obtain

$$v_i = \frac{v_S}{1 + j\omega R[C_{GS} + (1 - A_0)C_{GD}]}, \tag{3.43}$$

where $A_0 = -10$. Consequently,

$$v_o = \frac{A v_S}{1 + j\omega R[C_{GS} + (1 - A_0)C_{GD}]} \tag{3.44}$$

and, for a $1\,k\Omega$ signal source impedance, this will imply a bandwidth of $\omega_B = R^{-1}(C_{GS} + (1 - A_0)C_{GD})^{-1}$ or 8 MHz.

For broadband operation, we need to be able to overcome the Miller effect. One method of achieving this is the common-gate amplifier (an example is shown in Figure 3.42). Like its BJT counterpart (the common-base amplifier), the common-gate amplifier has a high gain ($\approx g_m R_D$) and is relatively immune from the Miller effect. It has a low input impedance ($\approx g_m^{-1} \| R_S$) and a high output impedance ($\approx R_D$). Another means of reducing the Miller effect is to use a cascode configuration, an example of which is shown in Figure 3.43. The same RF current i flows in both FET devices (they

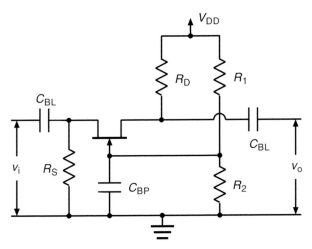

Figure 3.42   Common-gate amplifier.

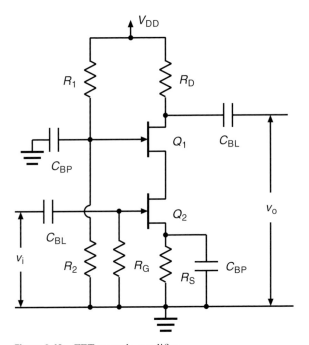

Figure 3.43   FET cascode amplifier.

are assumed to be identical) and so , since $v_o = -i R_D$ and $i = g_m v_i$, it follows that

$$\frac{v_o}{v_i} = -R_D g_m, \tag{3.45}$$

that is, the amplifier has a gain of $-R_D g_m$. The difference from the standard common-source amplifier is that the voltage gain of Q_2 is only -1, hence avoiding the Miller effect. This follows from the fact that minus the drain voltage fluctuations of Q_2 will

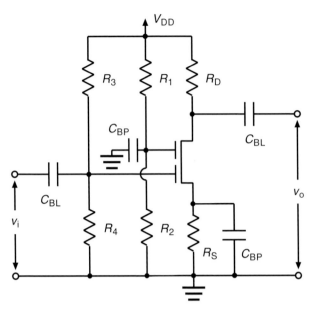

Figure 3.44   Dual-gate MOSFET amplifier.

be equal to the gate–source voltage fluctuations at Q_1. Since both transistors carry the same current, these fluctuations will be the same as the gate–source fluctuations of Q_2 and hence equal to v_i. Transistor Q_1 is essentially a common-gate amplifier and does not suffer from the Miller effect. The FET cascode configuration is so important that the dual-gate FET has been developed for its ease of implementation. This transistor is essentially two FET devices in series and can be modelled as such. In the dual-gate FET, however, the source of the top transistor connects continuously to the drain of the bottom transistor. An example of an amplifier using a dual-gate MOSFET is shown in Figure 3.44. It should be noted that the gain of the amplifier is controlled by the bias voltage on the upper gate of the transistor. Another amplifier that is immune from the Miller effect is the source follower amplifier (equivalent to the emitter follower amplifier), a typical example of which is shown in Figure 3.45. Like its BJT counterpart, this amplifier has a voltage gain slightly less than 1, a high input impedance and a low output impedance ($\approx g_m^{-1} \| R_S$).

FET differential amplifiers are important in that they have a very high input impedance and can form versatile building blocks for RF complementary MOS (CMOS) designs. A simple CMOS differential amplifier is shown in Figure 3.46. The analysis is similar to that for the BJT case and yields a differential gain of $-g_m R_D$ ($v_o^2 - v_o^1 = -g_m R_D(v_i^2 - v_i^1)$) and a differential output impedance of $2(R_D \| r_d)$ ($R_D \| r_d$ for each side). The bottom transistor is a current source with current I_{bias} set by the bias voltage V_{bias} (note that $I_{bias} = K(V_{bias} - V_t)^2$) and its high impedance will largely eliminate the common mode from the output. As with BJT designs, the

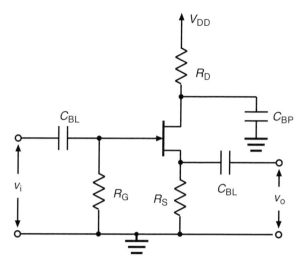

Figure 3.45 Source follower amplifier.

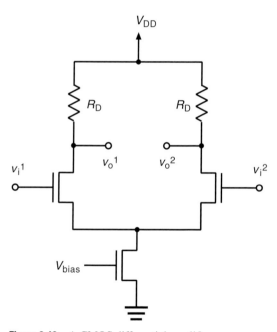

Figure 3.46 A CMOS differential amplifier.

drain resistors R_D can be replaced by active loads and some examples of these are shown in Figure 3.47. The first three loads have an impedance of value g_m^{-1} (fairly low). In the last three, however, the current is fixed by the gate to source bias voltage and so the loads will have an impedance equal to the drain resistance r_d of the device (a large value).

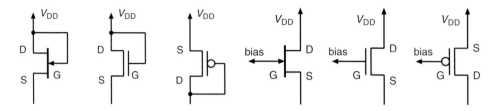

Figure 3.47 FET active loads.

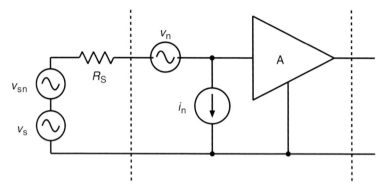

Figure 3.48 Amplifier with noise model.

3.8 Amplifier noise

The standard model of amplifier noise (Figure 3.48) consists of both current i_n and voltage v_n sources at the input to a noiseless amplifier. In addition, there will be a noise component v_{sn} contributed by the signal source. This will generate a noise power of $\overline{v_{sn}^2}/R_s = 4kTB$, where B is the bandwidth of the amplifier, R_s is the source resistance and T is the noise temperature of the source. The signal-to-noise ratio is given by

$$\frac{S}{N} = \frac{\frac{v_s^2}{R_s}}{\frac{\overline{v_n^2}}{R_s} + \overline{i_n^2}R_s + 4kTB},$$ (3.46)

where v_s is the signal voltage. Assuming the averaged signal power to be constant, the SNR will be maximum when $R_s^2 = \overline{v_n^2}/\overline{i_n^2}$. This source impedance, however, will not necessarily be same as that which gives the maximum power transfer. The conflicting requirements of noise and power match can sometimes be resolved by means of a process known as *emitter degeneration*. If an inductance L is added in series with the emitter of a BJT amplifier, this will cause an additional resistive component at the base. Since we have only added inductance to the circuit, the noise sources will be unchanged (pure reactance produces no noise). Consequently, it is possible to adjust the input impedance of the BJT without affecting the noise impedance and hence achieve simultaneous conjugate and noise match.

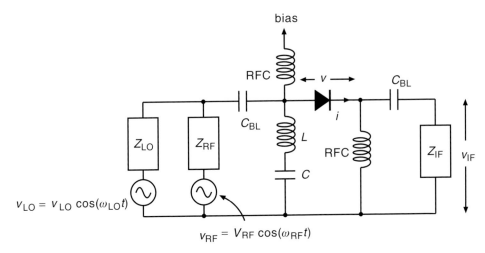

Figure 4.1   Diode mixer.

where v_D is the potential across the diode. For small voltage fluctuations v about the quiescent state, the relation can be expanded in a Taylor series to yield

$$i = I_0 + a_1 v + a_2 v^2 + \dots , \tag{4.2}$$

where I_0 is the quiescent current. Let $v_{RF} = V_{RF} \cos(\omega_{RF} t)$ and $v_{LO} = V_{LO} \cos(\omega_{LO} t)$ be the LO and RF voltages, respectively. The current in the diode will be given by

$$i = I_0 + a_1 (v_{LO} + v_{RF} - v_{IF}) + a_2 (v_{LO} + v_{RF} - v_{IF})^2 + \dots , \tag{4.3}$$

where $v_{IF} = V_{IF} \cos(\omega_{IF} t)$ is the IF voltage. Expanding the square term, we obtain

$$i \approx I_0 + a_1 (v_{LO} + v_{RF} - v_{IF}) + a_2 \left(v_{LO}^2 + v_{RF}^2 + v_{IF}^2 + 2 v_{LO} v_{RF} - 2 v_{LO} v_{IF} - 2 v_{RF} v_{IF} \right) \tag{4.4}$$

and hence, on noting that $2 \cos^2 A = 1 + \cos 2A$ and $2 \cos A \cos B = \cos(A - B) + \cos(A + B)$,

$$i \approx I_0 + a_1 (V_{LO} \cos \omega_{LO} t + V_{RF} \cos \omega_{RF} t - V_{IF} \cos \omega_{IF} t)$$
$$+ a_2 \frac{V_{LO}^2}{2} (1 + \cos 2\omega_{LO} t) + a_2 \frac{V_{RF}^2}{2} (1 + \cos 2\omega_{RF} t) + a_2 \frac{V_{IF}^2}{2} (1 + \cos 2\omega_{IF} t)$$
$$+ a_2 V_{LO} V_{RF} (\cos(\omega_{RF} - \omega_{LO}) t + \cos(\omega_{RF} + \omega_{LO}) t)$$
$$- a_2 V_{LO} V_{IF} (\cos(\omega_{LO} - \omega_{IF}) t + \cos(\omega_{LO} + \omega_{IF}) t)$$
$$- a_2 V_{RF} V_{IF} (\cos(\omega_{RF} - \omega_{IF}) t + \cos(\omega_{RF} + \omega_{IF}) t). \tag{4.5}$$

From this last expression, the component of current at the intermediate frequency $\omega_{IF} = |\omega_{RF} - \omega_{LO}|$ has amplitude

$$I_{IF} = -a_1 V_{IF} + a_2 V_{LO} V_{RF}. \tag{4.6}$$

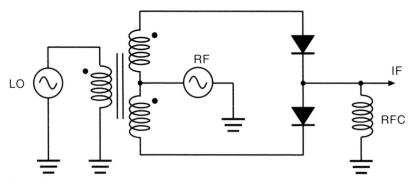

Figure 4.2 Balanced mixer based on a diode pair.

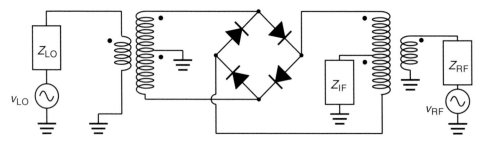

Figure 4.3 Diode ring mixer.

Since $V_{IF} = I_{IF} Z_{IF}$, it follows that

$$V_{IF} = \frac{Z_{IF} a_2 V_{LO} V_{RF}}{1 + Z_{IF} a_1}. \tag{4.7}$$

It will be noted that the *voltage conversion gain* of the mixer $Z_{IF} a_2 V_{LO}/(1 + Z_{IF} a_1)$ is dependent upon the level of the local oscillator signal.

A single diode mixer has the disadvantage that the output will contain strong contributions from both the RF input and LO signals. Consequently, there will need to be suitable filtering at the IF load Z_{IF}. A partial solution can be obtained by adopting the two diode mixer shown in Figure 4.2. The diodes on either side of the circuit will conduct currents $I_s \exp[(v_{LO} + v_{RF} - v_{IF})/V_T]$ and $-I_s \exp[(v_{LO} - v_{RF} - v_{IF})/V_T]$, respectively. Together, these currents cause an IF voltage that has a component proportional to the product of the RF and LO signals, but with no component at the LO frequency. Besides the desired components at the sum and difference frequencies, however, the IF output will also contain a component at the RF frequency. As a consequence, the IF load will require filtering to remove any content at frequency ω_{RF}. An alternative is the *diode ring* mixer (see Figure 4.3) which exhibits a very high degree of isolation (the IF output has very little signal at other than the sum and difference frequencies $|\omega_{LO} \pm \omega_{RF}|$). Unfortunately, there is a downside in that the conversion gain is very low (typically about -7 dB). The mixer works through the switching action of the diode bridge. As

the BJT. Consequently, as far as mixing is concerned, a long-channel FET is likely to provide a far superior result.

4.3   Transconductance mixers

The gain of a differential amplifier depends on the transconductance g_m of its transistors and this transconductance will, in turn, depend on the current flow. Consequently, a differential amplifier can be transformed into a mixer if it is biased by a current source that is controlled by one of the signals to be mixed (usually the low-level RF signal). Figure 4.8 shows such a mixer (bias components not included). If v_{RF} denotes the RF part of the current source base voltage, this transistor will exhibit a collector current

$$i_C \approx i_{RF} + I_E \approx I_E \exp\left(\frac{v_{RF} - R_E i_{RF}}{V_T}\right) \approx I_E\left(1 + \frac{v_{RF} - R_E i_{RF}}{V_T}\right), \qquad (4.14)$$

where I_E is the d.c. component of emitter current and i_{RF} is the RF component. We will thus have $i_{RF} \approx I_E v_{RF}/(V_T + R_E I_E)$ and, noting that $V_T \ll I_E R_E$ under normal circumstances, $v_{RF} \approx R_E i_{RF}$. It is clear from this that we can approximate the collector current by $i_C \approx I_E + v_{RF}/R_E$. The output of the differential amplifier is given by

$$v_{IF} = -g_m R_L v_{LO}, \qquad (4.15)$$

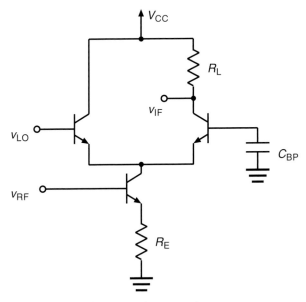

Figure 4.8   A basic transconductance mixer.

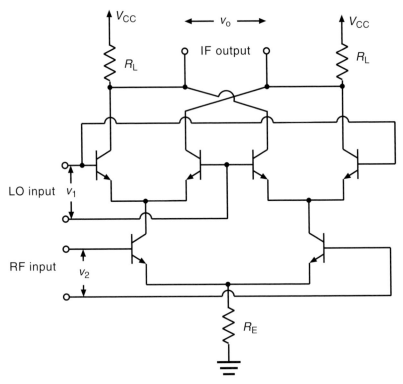

Figure 4.9   Gilbert cell mixer.

where $g_m = i_C/2V_T$ (note that current i_C is split between the two transistors of the differential pair). Consequently,

$$v_{IF} = -\left(\frac{R_L I_E}{2V_T}\right) v_{LO} \left(1 + \frac{v_{RF}}{R_E I_E}\right) \tag{4.16}$$

and the output will have amplitude $R_L V_{LO} V_{RF}/4R_E V_T$ at the IF frequency $|\omega_{LO} - \omega_{RF}|$. Note, however, that the output will also contain a strong component at the local oscillator frequency. This component can be removed by putting together two transconductance mixers in a fashion that cancels the unwanted component. Such a mixer is known as a Gilbert cell and an example (bias components not shown) is given in Figure 4.9.

Transconductance mixers can be constructed equally well using FET differential amplifiers and suitable designs are left to the reader as an exercise. The transconductance mixer is an example of a single balanced mixer (either the LO signal or the RF signal is suppressed at the IF output) whereas the diode ring and Gilbert cell mixers are examples of double balanced mixers (both the LO an RF signals are suppressed at the IF output). The choice of mixer depends on the degree of isolation that is required.

4.4   Amplitude modulation

Thus far, we have concentrated upon RF signals with pure sinusoidal time dependence. In reality, however, an RF signal will be required to convey information and, as we have seen earlier, this will necessitate some form of modulation on the sinusoidal *carrier* wave. In the case of amplitude modulation (AM), the modulated signal will have the form

$$v(t) = [1 + m_{\mathrm{I}} a(t)] A_{\mathrm{c}} \cos(\omega_{\mathrm{c}} t), \qquad (4.17)$$

where ω_{c} is the carrier frequency, $|a(t)| < 1$ and m_{I} is the modulation index (usually expressed as a percentage with $m_{\mathrm{I}} = 1$ corresponding to 100%). The function $a(t)$ represents the *baseband* signal that contains the information to be conveyed. A transconductance mixer can be used to achieve amplitude modulation. If we replace the RF input v_{RF} by the baseband signal v_{AF} (AF because this is often an audio signal), the IF output will be

$$v_{\mathrm{IF}} = -\left(\frac{R_{\mathrm{L}} I_{\mathrm{E}}}{2 V_{\mathrm{T}}}\right) v_{\mathrm{LO}} \left(1 + \frac{v_{\mathrm{AF}}}{R_{\mathrm{E}} I_{\mathrm{E}}}\right). \qquad (4.18)$$

For the sinusoidal baseband signal $v_{\mathrm{AF}}(t) = V_{\mathrm{AF}} \sin \omega_{\mathrm{AF}}(t)$, the modulation index will be $m_{\mathrm{I}} = V_{\mathrm{AF}}/R_{\mathrm{E}} I_{\mathrm{E}}$.

A simple AM transmitter will take the form shown in Figure 4.10. With such a system, however, the power amplifier (PA) will need to be linear in order to avoid distorting the modulation. Otherwise, the modulation will need to be kept at a low level (m_{I} much less than 1). For this reason, the RF signal is normally taken to full power before the modulation is applied. This allows a highly efficient class C amplifier to be used in the final stage of the transmitter. The downside, however, is that the modulating signal must also be taken to a high power level.

The simplest variety of AM demodulator is the envelope detector, an example of which is shown in Figure 4.11. Basically, the detector consists of a diode rectifier followed by a low-pass filter. The filter will need to have large enough bandwidth ω_{B} to pass the modulating baseband signal, but small enough bandwidth to remove

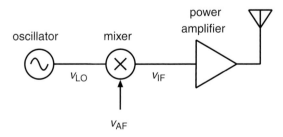

Figure 4.10   Simple AM transmitter.

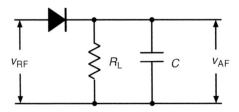

Figure 4.11 Simple AM detector.

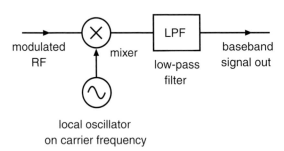

Figure 4.12 Synchronous detector.

any residual carrier. To achieve this, R_L and C will need to be chosen such that $R_L C \gg 2/\omega_c$ and $R_L C < 2/\omega_B$ (note that R_L represents the load presented by subsequent stages of the receiver). The output of the detector will be proportional to $1 + m_I a(t)$ which has an a.c. component that is proportional to the original modulating signal.

Another technique for extracting the baseband signal is known as synchronous demodulation. In this approach, the modulated RF signal is mixed with a signal at the carrier frequency (see Figure 4.12). For incoming signal v, the effect of the mixer will be

$$v(t)\cos(\omega_c t) = [1 + m_I a(t)]A_c \cos(\omega_c t + \phi)\cos(\omega_c t)$$
$$= \frac{A_c}{2}[1 + m_I a(t)](\cos(2\omega_c t + \phi) + \cos \phi), \qquad (4.19)$$

where ϕ takes account of the phase difference between the incoming carrier and the local oscillator signal. The low-pass filter after the mixer is designed to remove all but the baseband signal. Consequently, the output of the detector will be $(A_c/2)[1 + m_I a(t)]\cos \phi$ which has an a.c. component that is proportional to the original modulating signal.

An AM modulated signal contains a considerable amount of redundant information in that it consists of a carrier signal (frequency ω_c), a copy of the baseband signal translated upwards in frequency by an amount ω_c (the *upper sideband*) and a copy of the upper sideband signal reflected in the carrier frequency (the *lower sideband*). This is illustrated in Figure 4.13. Consider the sinusoidal modulating signal $a(t) = \cos(\omega_{AF} t)$,

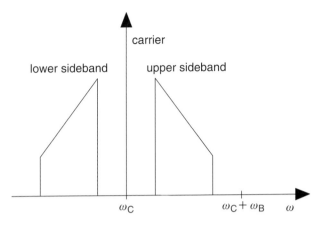

Figure 4.13   Sideband structure.

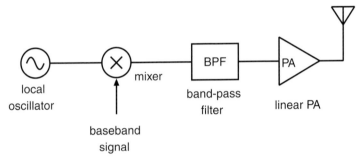

Figure 4.14   Simple SSB transmitter.

then

$$v(t) = A_c[1 + m_I \cos(\omega_{AF}t)] \cos(\omega_c t)$$

$$= A_c \cos(\omega_c t) + \frac{m_I A_c}{2} \cos[(\omega_c + \omega_{AF})t] + \frac{m_I A_c}{2} \cos[(\omega_c - \omega_{AF}t)]. \ (4.20)$$

In the above expansion, the first term is the carrier component, the second term the upper sideband and the third term the lower sideband. It is clear that the full AM signal will have a bandwidth twice that of the baseband signal (bandwidth ω_B). This is wasteful as the carrier conveys no information and the two sidebands contain exactly the same information. In reality, we only need to transmit one of the sidebands. This single sideband (SSB) mode halves the required bandwidth of the transmitted signal and considerably reduces the power handling requirements of the transmitter power amplifier stages. The PA, however, will need to exhibit a high degree of linearity.

It is possible to produce an SSB signal by suitably filtering a low-level AM modulated signal. The signal level is then raised by means of a linear power amplifier. A block diagram of such a transmitter is shown in Figure 4.14. The baseband signal (bandwidth ω_B) is mixed with a local oscillator signal at a frequency of ω_c to produce a signal with both sidebands. (Normally, the mixer will be of the balanced variety in order to remove

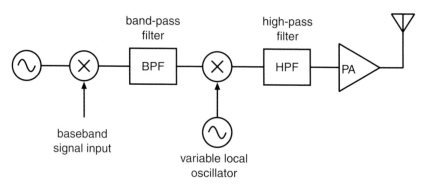

Figure 4.15   Variable frequency SSB transmitter.

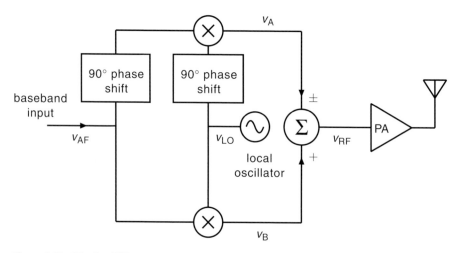

Figure 4.16   Phasing SSB generation.

contributions at the local oscillator frequency.) This signal is then passed through a band-pass filter (BPF) with edge frequencies $\omega_c + \omega_\delta$ and $\omega_c + \omega_B$ when the upper sideband is required and $\omega_c - \omega_B$ and $\omega_c - \omega_\delta$ when the lower sideband is required (ω_δ is a small frequency offset to guard against residual local oscillator contribution).

   If a transmitter is required to cover a variety of frequencies, this is usually achieved through a second stage of mixing due to the difficulty of producing a variable frequency band-pass filter with sufficient performance. A typical transmitter architecture is illustrated in Figure 4.15. A signal containing both sidebands is generated at a fixed frequency and band-pass filtered to select the desired sideband. The output is then mixed with the signal from a variable frequency oscillator and the sum frequency selected by means of a high-pass filter (HPF). To accommodate the variations in frequency, the PA will need to be of the broadband variety.

   An alternative technique for generating SSB modulation is the phasing method (a block diagram of the generator is shown in Figure 4.16). The operation of the phasing method is most easily understood by considering a single frequency baseband signal.

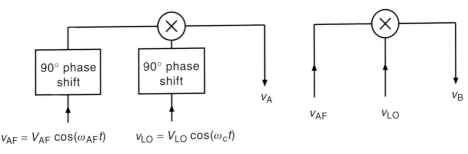

Figure 4.17   Analysis of the phasing technique.

We can view the modulator as consisting of two mixers, each fed by baseband and local oscillator signals. For one of the mixers, however, both input signals need to have their phase advanced by 90°. The output of these mixers will consist of

$$
v_A = V_{AF} V_{LO} \cos\left(\omega_c t + \frac{\pi}{2}\right) \cos\left(\omega_{AF} t + \frac{\pi}{2}\right)
$$

$$
= \frac{V_{AF} V_{LO}}{2} \{\cos[(\omega_c - \omega_{AF})t] - \cos[(\omega_c + \omega_{AF})t]\} \tag{4.21}
$$

and

$$
v_B = V_{AF} V_{LO} \cos(\omega_c t) \cos(\omega_{AF} t)
$$

$$
= \frac{V_{AF} V_{LO}}{2} \{\cos[(\omega_c - \omega_{AF})t] + \cos[(\omega_c + \omega_{AF})t]\}. \tag{4.22}
$$

When the mixer outputs are summed

$$
v_{RF} = v_A + v_B = V_{AF} V_{LO} \cos[(\omega_c - \omega_{AF})t] \tag{4.23}
$$

which is the lower sideband. The upper sideband results when we take the difference between the mixer outputs.

The synchronous demodulator of Figure 4.12 can also be used to detect single sideband signals. Unfortunately, besides the desired sideband, this system will also attempt to demodulate any signal that occupies the frequencies that are associated with the complementary sideband. To overcome this, a detector can be constructed using the phasing principle (a block design is shown in Figure 4.18). The phasing demodulator has the effect of cancelling out the contribution from a signal that occupies the complementary sideband.

The key to phasing techniques is the ability to perform a 90° phase shift on both local oscillator and baseband signals. This can be achieved with RC networks in the case of narrow bandwidth signals (Figure 4.19 shows a narrow band phase shifter). It should be noted, however, that the baseband signal will often require an accurate 90° phase shift over a wide range of frequencies and this can be difficult to achieve with RC circuits. Modern systems, however, achieve the phase change through digital techniques.

A more traditional means of demodulating single sideband signals is through a super-heterodyne receiver (an example is shown in Figure 4.20). By suitable choice of local

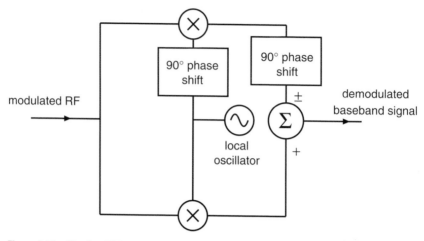

Figure 4.18 Phasing SSB demodulation.

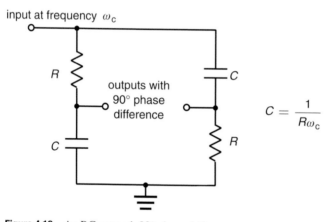

Figure 4.19 An RC network 90° phase shifter.

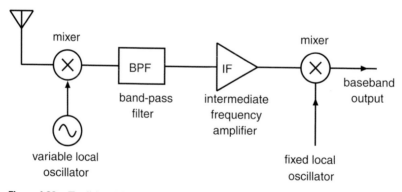

Figure 4.20 Traditional SSB receiver.

oscillator frequency, the desired signal is reduced to a frequency range that can pass through a fixed frequency band-pass filter (bandwidth equal to the bandwidth ω_B of the original baseband signal). Only one sideband will pass through the filter and this can be demodulated by a standard synchronous detector with a fixed frequency local oscillator.

4.5   Angle modulation

In angular modulation, information is transferred as variations in phase (phase modulation) or rate of change of phase (frequency modulation). The RF signal takes the form

$$v(t) = A_c \cos(\omega_c t + \phi), \tag{4.24}$$

where $\phi = ma(t)$ corresponds to phase modulation (PM) and $\phi = m \int a(t)\,dt$ corresponds to frequency modulation (FM). For the sinusoidal modulating signal $a(t) = \sin(\omega_{AF} t)$, phase modulation will result in the signal

$$v(t) = A_c \cos(\omega_c t + m \sin(\omega_{AF} t)) \tag{4.25}$$

and frequency modulation in the signal

$$v(t) = A_c \cos\left[\omega_c t - \frac{m}{\omega_{AF}} \cos(\omega_{AF} t)\right]. \tag{4.26}$$

The *modulation index* β is defined to be the maximum deviation of ϕ. For phase modulation, this is m and for frequency modulation m/ω_{AF}. It can be shown that a carrier that is angle modulated by a sinusoid has bandwidth $2\omega_{AF}(\beta + 1)$ and from this it is clear that angle modulated signals occupy more bandwidth than their amplitude modulated counterparts. Angle modulated signals do, however, have much greater immunity to interference if properly demodulated.

A simple technique for producing FM is to control a VCO with a suitably scaled version of the baseband signal. Phase modulation is a little more difficult, but can be produced using the circuit shown in Figure 4.21. The FET will act as a mixer to produce an output voltage $k_1 v_{AF} v_{RF}$ where k_1 is a constant depending on the quiescent operating conditions. In addition, it will act as a feedback amplifier in which the capacitive feedback will cause a $90°$ phase change in the RF signal ($V_{RF} \cos(\omega_c t)$ is transformed into $k_2 V_{RF} \cos(\omega_c t + \pi/2)$, where k_2 is a constant that depends on the quiescent conditions). The combination of these two outputs (all other outputs are removed by the filtering action of subsequent circuits) is equivalent to a phase modulated copy of the input RF. Providing that $k_1 V_{AF}/k_2$ is suitably small, the output $k_2 V_{RF} \cos(\omega_c t + \pi/2) + k_1 v_{AF} V_{RF} \cos(\omega_c t)$ is approximately equal to the phase modulated carrier $k_2 V_{RF} \cos[\omega_c t + \pi/2 - (k_1/k_2)v_{AF}]$ on noting that

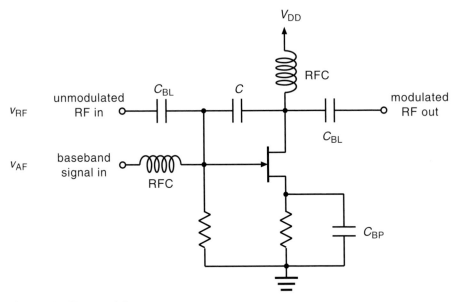

Figure 4.21   Phase modulator.

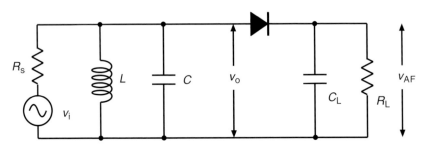

Figure 4.22   Simple FM demodulation.

$\cos[\theta + \pi/2 + \delta\theta] \approx \cos(\theta + \pi/2) - \delta\theta \cos\theta$. By frequency tailoring the baseband input, it is also possible to use such a phase modulator to generate narrow band FM (NBFM) signals.

The frequency dependence of a tuned circuit can be utilised for the demodulation of an FM signal. If the resonance frequency ω_0 is offset from the centre frequency ω_c of the FM signal, the frequency variations of the FM will result in an amplitude modulated FM signal that can be demodulated by standard AM techniques. A suitable circuit is shown in Figure 4.22, for which

$$|v_o| = \frac{R}{(R + R_s)\left[1 + \frac{4Q^2}{\omega_0^2}(\omega - \omega_0)^2\right]^{1/2}}|v_i|, \qquad (4.27)$$

where $Q = (R_s \parallel R)/\omega_0 L$ and R represents the total load presented by the envelope detector. For an FM signal, all the information is carried as frequency deviations and

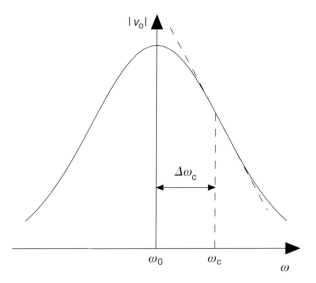

Figure 4.23   Detector frequency response.

so $|v_i|$ is essentially constant (this situation is usually guaranteed by means of a limiting amplifier that precedes the demodulator). As a consequence, $|v_o|$ will respond to frequency deviations in the fashion of Figure 4.23. If the frequency offset $\Delta\omega_c$ is suitably chosen, the amplitude of $|v_o|$ will vary linearly with the frequency variation about ω_c. This will result in an amplitude modulated FM signal that is demodulated by the envelope detector. Phase demodulation is a little more difficult, but it is possible to use a synchronous detector with local oscillator set to the centre frequency of the modulated signal. For small phase deviations, the a.c. component of the output will be proportional to the baseband signal.

4.6   Gain and amplitude control

The transconductance mixer described earlier in this chapter can perform the function of an amplifier with voltage controlled gain, the control voltage being applied to the gate/base of the device that controls the source/emitter current of the differential pair. Amplifiers with electronically controlled gain are extremely important elements in a communication system. In many communication scenarios, especially in mobile communications, the signal strength varies continually and this can result in annoying amplitude variations in the demodulated signal. Normally, however, these variations will occur on a timescale that is much longer than the scales associated with the baseband signal. Consequently, if the amplitude of the demodulated output is averaged over a suitable time period (i.e., low-pass filtered), this voltage can be fed back to control the gain of the IF and/or RF amplifiers and hence compensate for fluctuations in signal

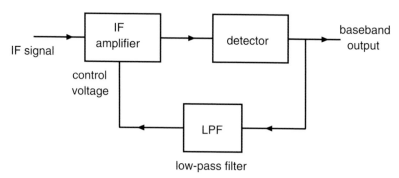

Figure 4.24   Automatic gain control.

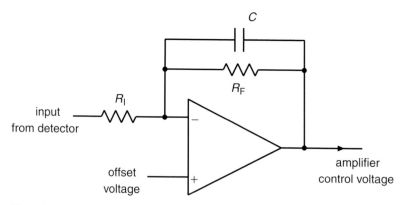

Figure 4.25   Filter and amplifier for AGC.

strength. The feedback needs to be engineered so that the amplifier gain falls and rises as the overall signal strength rises and falls. Such a system is known as automatic gain control (AGC) and an example, suitable for an AM receiver, is shown in Figure 4.24. Figure 4.25 shows a low-pass filter that is suitable for use in an AGC feedback circuit. It consists of a standard operational amplifier configuration with frequency dependent feedback in which the components R_I, R_F and C are used to tailor the gain and bandwidth (gain $= R_F/R_I(j\omega R_F C + 1)$).

In the case of FM signals, the amplitude does not carry any of the desired information and, as a consequence, an FM receiver will usually bring the amplitude up to a standard level. This will remove amplitude interference (static, ignition noise, etc.) from the FM signal and make possible very high quality reception. The uniform amplitude level is usually achieved through a limiting amplifier, an example of which is shown in Figure 4.26. The design is based upon an operational amplifier for which the gain is severely reduced when the output voltage is large enough to force the diodes into conduction. R_I and R_F together determine the small signal gain (a large value in order to bring up the lowest levels of signal) whilst R_I and R_L determine the large signal gain.

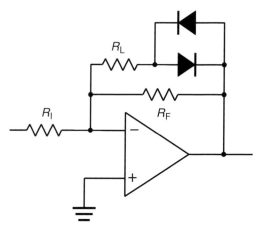

Figure 4.26   Limiting amplifier.

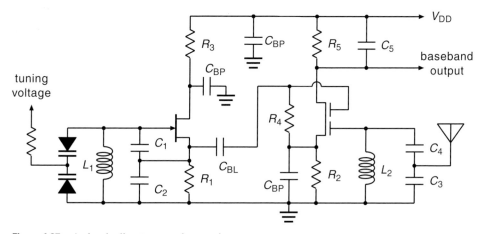

Figure 4.27   A simple direct conversion receiver.

4.7   A simple receiver design

If a synchronous detector is fed directly from the antenna (usually via a filter), it forms what is known as a direct conversion (DC) receiver (also known as a homodyne receiver). Figure 4.27 shows the circuit of a simple DC receiver. This receiver is designed around a single dual-gate FET mixer and a Colpitts oscillator that uses a JFET source follower amplifier (see Chapter 5). The baseband output is taken from the mixer drain and there is a capacitor across the drain resistor in order to form a low-pass filter (note that a realistic design will require a much greater degree of baseband filtering). The input circuit is resonated at the centre frequency ω_0 of the required reception band with Q chosen to produce the required bandwidth. Since $Q \approx R/\omega_0 L_2$ where R is the antenna resistance when transformed through the capacitive divider, the ratio of C_3 to C_4 will

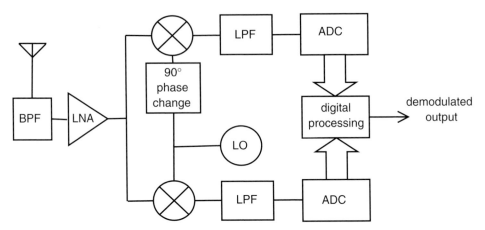

Figure 4.28   Modern form of DC receiver.

need to be chosen according to bandwidth. The oscillator frequency can be varied, and hence the receiver tuned, by means of the back-to-back varactor diodes across the oscillator inductor.

A DC receiver is arguably one of the simplest that can be constructed for the demodulation of AM and SSB. Its problem, however, is that it tries to demodulate signals either side of the oscillator frequency and is thus highly susceptible to co-channel interference in the case of SSB. Consequently, in today's crowded spectrum, the above form of DC receiver is not a serious option. We could, however, combine two DC receivers in the manner of a phasing demodulator and produce a receiver that overcomes the above drawback (see Figure 4.28). In this form, the DC receiver is becoming increasingly popular in communication systems. Each mixer is normally followed by an ADC (analogue-to-digital converter) with phase shift and demodulation performed digitally. The resulting receiver is extremely flexible and can handle virtually any form of modulation.

EXERCISES

(1) A single diode mixer is biased for 1 mA quiescent current flow ($I_s = 10^{-15}$ A and $V_T = 25$ mV). Calculate the conversion gain when the IF stage presents a 200 Ω load and there is a 10 mV local oscillator drive.

(2) Design a BJT mixer with conversion gain of 10 dB for an IF load of 1 kΩ and a local oscillator amplitude of 10 mV. (Calculate the bias components for a V_{CC} of 10 V.)

(3) Design a transconductance mixer with a conversion gain of 20 dB for a local oscillator drive of 10 mV and an IF load of 2 kΩ. (Calculate bias components for a V_{CC} of 10 V.)

(4) For the envelope detector shown in Figure 4.11, and a load R_L of $1\,k\Omega$, calculate a value of C that can achieve suitable filtering for a $3\,kHz$ bandwidth baseband signal. Assuming the diode to be sufficiently represented by the first three terms in the Taylor series expansion 4.2, calculate the harmonic distortion for an input voltage of $5\,mV$.

(5) An SSB signal can be demodulated by an envelope detector (see Figure 4.11) providing some carrier signal (frequency ω_c) has been reinserted before detection ($s_{REC} = A_c \cos(\omega_c t) + s_{SSB}(t)$ is the augmented signal before detection). By considering the single tone signal $s_{SSB}(t) = A\cos(\omega_c + \omega_{AF})t$ (baseband frequency ω_{AF}), and assuming the diode characteristic to be sufficiently approximated by $i(v) = (I_s/V_T)[v + v^2/2V_T]$ in forward bias, find the component at frequency ω_{AF} after detection and hence show that the original modulating signal can be recovered.

(6) Calculate component values for a DC receiver based on Figure 4.27 that will tune to your favourite AM broadcast station. Perform a simulation using a suitable circuit simulation package (you may use two series JFETs instead of the dual-gate MOSFET). Discuss the drawbacks of this simple design and suggest some remedies.

SOURCES

Everard, J. 2001. *Fundamentals of RF Circuit Design with Low Noise Oscillators*. Chichester: John Wiley.

Hayward, W. 1994. *Introduction to Radio Frequency Design*. Newark, CT: American Radio Relay League.

Lee, H. 1985. *The Design of RF CMOS Radio Frequency Integrated Circuits*. Cambridge University Press.

Ludwig, R. and Bretchko, P. 1990. *RF Circuit Design: Theory and Application*. Upper Saddle River, NJ: Prentice-Hall.

Maas, S. A. 1997. *Non-linear Microwave Circuits*. Piscataway, NJ: IEEE Press.

Pozar, D. M. 1998. *Microwave Engineering* (2nd edn). New York: John Wiley.

Pozar, D. M. 2001. *Microwave and RF Design of Wireless Systems*. New York: John Wiley.

Proakis, J. G. and Salehi, M. 1994. *Communication Systems Engineering*. Englewood Cliffs, NJ: Prentice-Hall.

Rhode, U. L., Whitaker, J. and Bucher, T. T. N. 1996. *Communication Receivers* (2nd edn). New York: McGraw-Hill.

Sabin, W. E. and Schoenike, E. O. (eds.). 1998. *HF Radio Systems and Circuits*. Atlanta, GA: Noble Publishing Corporation.

Smith, J. R. 1997. *Modern Communication Circuits*. New York: McGraw-Hill.

5 Oscillators and phase locked loops

The generation of a stable sinusoidal signal is a crucial function in most RF systems. A transmitter will amplify and suitably modulate such a signal in order to produce its required output. In the case of a receiver system, such a signal is fed into the mixer circuits for the purposes of frequency conversion and demodulation. A circuit that generates a repetitive waveform is known as an *oscillator*. Such circuits usually consist of an amplifier with positive feedback that causes any input, however small, to grow until limited by the non-linearities of the circuit. The feedback will need to be frequency selective in order to control the rate of waveform repetition. This frequency selection is often achieved using combinations of capacitors and inductors, but can also be achieved with resistor and capacitor combinations. In the present chapter, however, we will concentrate on feedback circuits based on capacitor/inductor combinations. We consider a variety of oscillator circuits that are suitable for RF purposes and investigate the conditions under which oscillation occurs. In addition, we consider the issue of oscillator noise since this can often pose a severe limitation upon system performance.

A particularly important class of oscillator is that for which the frequency can be controlled by a d.c. voltage. Such an oscillator is an important element in what is known as a *phase locked loop*. In such a system, there is a feedback loop that compares the oscillator output with a reference signal and generates a control voltage based upon their phase difference. When the system settles down, the oscillator is *locked* onto the reference signal. Phase locked loops form a generic class of system that can been used for purposes such as frequency control and demodulation. Here we consider the basic principles of phase locked loops and investigate some of their applications.

5.1 Feedback

A general amplifier (H) with feedback (G) is illustrated in Figure 5.1. For this system, the relationship between input and output voltages is given by

$$v_{o} = \frac{H(j\omega)}{1 \pm G(j\omega)H(j\omega)} v_{i}.$$

(5.1)

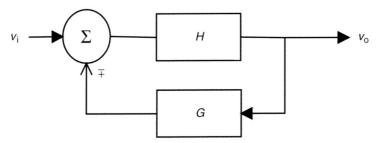

Figure 5.1 General feedback system.

Negative feedback (the positive sign in Equation 5.1) provides a means of tailoring the amplifier frequency response and controlling both gain and linearity. When there is positive feedback, however, the system can provide a means of generating oscillations. For oscillation, we require a system that produces a signal without input (except for an initial excitation). Consequently, for oscillations to occur at frequency ω_0,

$$|G(j\omega_0)||H(j\omega_0)| = 1 \tag{5.2}$$

and

$$\arg\{G(j\omega_0)H(j\omega_0)\} = 0. \tag{5.3}$$

This is the Barkhausen criterion and it ensures that any small component at frequency ω_0 will grow until limited by the non-linearities of the system.

5.2   The Colpitts oscillator

In a Colpitts oscillator, the feedback occurs via a series inductance π-network. An example, based on a common-emitter BJT amplifier, is shown in the circuit of Figure 5.2. By neglecting all but the current source of the BJT ($r_\pi = r_0 = \infty$ and $C_\mu = C_\pi = 0$), we obtain the simplified small signal model of Figure 5.3. Current balance at the BJT collector will imply

$$sC_2 v_\pi + g_m v_\pi + \left(\frac{1}{R} + sC_1\right) v_0 = 0 \tag{5.4}$$

and, at the base,

$$sC_2 v_\pi = \frac{v_0 - v_\pi}{sL}. \tag{5.5}$$

From Equation 5.5 we obtain $v_0 = v_\pi(1 + s^2 C_2 L)$ and then, eliminating v_0 from Equation 5.4,

$$sC_2 + g_m + \left(\frac{1}{R} + sC_1\right)(1 + s^2 C_2 L) = 0. \tag{5.6}$$

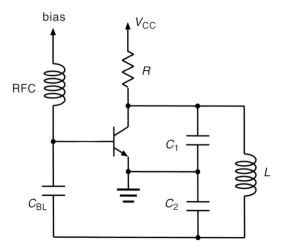

Figure 5.2   Colpitts oscillator.

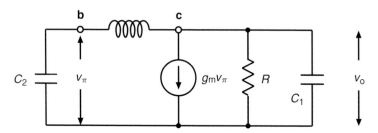

Figure 5.3   Simple model of Colpitts oscillator.

Noting that $s = j\omega$, and separating the real and imaginary parts of Equation 5.6, we obtain

$$\omega = \sqrt{\frac{C_1 + C_2}{LC_1C_2}} \quad \text{and} \quad \frac{C_2}{C_1} = g_m R, \tag{5.7}$$

where ω is the frequency of oscillation, C_2/C_1 is the feedback ratio and $g_m R$ is the voltage gain of the amplifier. A practical design will normally set the transistor gain slightly higher than the feedback ratio ($C_2/C_1 < g_m R$) to take account of component variations. In this case, the oscillations will grow until the non-linearities in the device cause sufficient loss of gain for the Barkhausen criterion to be satisfied. This last point is important as it means that the steady state operation of an oscillator is essentially non-linear. If v_π is large, the transistor will make excursions into regions where it is switched off (we usually set the bias so that this is the case). Consequently, the transistor collector current I_t will consist of periodic pulses with the peaks occurring where v_π is maximum. The bias current I_{bias} will be the average of the transistor current pulses ($I_{\text{bias}} = (1/T) \int_0^T I_t \, dt$, where $T = 2\pi/\omega$) and, by Fourier techniques, the fundamental component of the current I_{fund} is given by $I_{\text{fund}} = (2/T) \int_0^T I_t \cos(\omega t) \, dt$. The major contribution to I_{fund} will arise around the peak of v_π and, as a consequence, $I_{\text{fund}} \approx 2I_{\text{bias}}$. If $v_\pi = V_{\text{osc}} \cos(\omega t)$

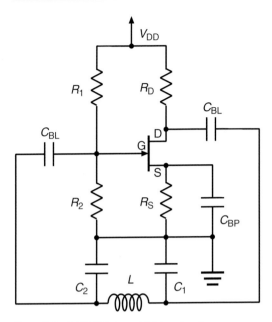

Figure 5.4   FET Colpitts with bias circuits.

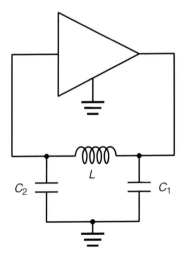

Figure 5.5   General Colpitts oscillator.

(feedback only occurs at the fundamental frequency) there will be an effective large sig-nal transconductance of $G_m = 2I_{bias}/V_{osc}$ (see Lee). This is a general relationship, but for a BJT it reduces to $G_m = (2V_T/V_{osc})g_m$ and for an FET to $G_m = [(V_{GS} - V_t)/V_{osc}]g_m$, where V_{GS} is the d.c. component of the gate–source voltage. In designing an oscillator, we will need to ensure that the Barkhaussen criterion can be satisfied somewhere between the extremes of the large and small signal transconductances.

An FET version of the Colpitts oscillator is shown in Figure 5.4. If we replace the FET amplifier by a generic unit, we obtain the generic Colpitts oscillator shown in Figure 5.5.

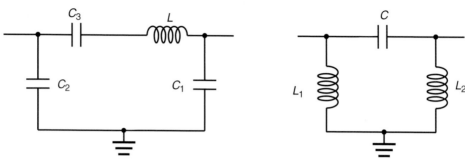

Figure 5.6   Clapp and Hartley feedback circuits.

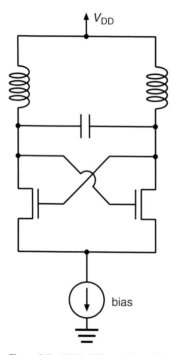

Figure 5.7   FET differential oscillator.

(Note that the feedback requires the amplifier to have a phase shift of 180°, which is the case for common-source and common-emitter amplifiers.) The Colpitts circuit employs a series inductance π-network feedback, but there are alternative feedback circuits that give rise to the Clapp and Hartley oscillators (see Figure 5.6). Whilst oscillators based on a single-ended amplifier input are common, it is also possible to base an oscillator on a differential amplifier. Figure 5.7 shows a design that is suitable for CMOS implementation. Positive feedback is achieved by using both differential input and output, the output of one side feeding the input of the other.

In practical applications, the oscillator will need to act as a source of RF signals. Consequently, this will mean an additional load on the circuit. The signal is normally taken from the output of the amplifier on which the oscillator is based (the collector in

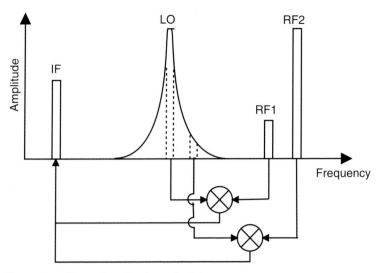

Figure 5.15   Illustration of reciprocal mixing.

where v_{sig} is the noise free signal level. Phase noise imposes an important limitation of oscillator performance and oscillator specification will normally include values of relative phase noise at various frequency offsets from the intended frequency. In particular, for a receiver that down converts the input RF signal to an intermediate frequency (IF), the phase noise performance of the receiver *local oscillator* (LO) can be important because of the possibility of *reciprocal mixing*. Although the desired signal will mix with the intended LO frequency to produce the IF frequency, there is the possibility that energy from the local oscillator frequency *skirts* could mix with strong out-of-band signals to also produce the IF frequency. This is known as reciprocal mixing and will lead to interference that could be unacceptable in some applications. The concept is illustrated in Figure 5.15 where RF1 represents the desired signal and RF2 the strong out of band signal.

It is clear that we require high Q resonant circuits for good oscillator performance. Extremely high Q resonators can be constructed out of quartz crystals and these are used extensively in RF circuits. Such resonators are electromechanical in nature and use the piezoelectric effect to translate high quality mechanical vibrations into electrical oscillations. (Other devices with high Q are coaxial and ceramic resonators.) Figure 5.16 shows a typical circuit model for a quartz crystal (valid near the fundamental frequency of resonance). The inductance is extremely large (hundreds of henries) and the shunt capacitance C_2 is typically tens of picofarads (the series capacitance C_1 is very much less). From the circuit model, we obtain the following expression for the impedance of the crystal

$$Z = \frac{(1 - \omega^2 L C_1) + j\omega R C_1}{j\omega(C_1 + C_2)\left(1 - \frac{\omega^2 L C_1 C_2}{C_1 + C_2}\right) - \omega^2 R C_1 C_2}. \tag{5.23}$$

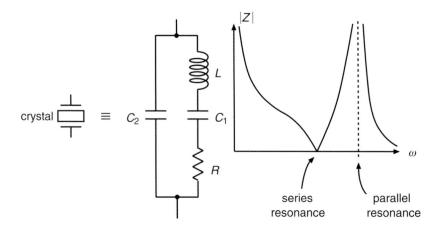

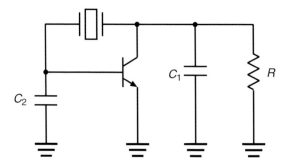

Figure 5.16   Crystal resonator model.

Figure 5.17   Crystal Colpitts oscillator.

It is clear that the device will exhibit both series and parallel resonance (note that R can be neglected due to the very high Q). The frequencies of these resonances will, however, be very close. Figure 5.17 shows a typical example of a crystal controlled Colpitts oscillator (bias components and d.c. supply not shown).

5.4   Voltage controlled oscillators

An oscillator with a voltage controlled frequency is often required in applications such as phase locked loops. A varicap (variable capacitance) diode (sometimes known as a varactor) can be used to achieve this. Diodes can be manufactured such that the reverse bias junction capacitance changes quite dramatically with voltage

$$C(V_{\text{bias}}) = \frac{C_0}{\left(1 - \frac{V_{\text{bias}}}{V_{\text{diff}}}\right)^{\frac{1}{2}}}, \qquad (5.24)$$

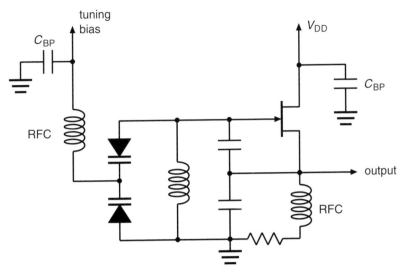

Figure 5.18 A Colpitts VCO.

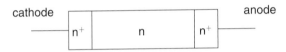

Figure 5.19 Gunn diode.

where C_0 is the zero bias capacitance and V_{diff} has a value of about 0.6 V for a silicon diode. Figure 5.18 shows a Colpitts voltage controlled oscillator (VCO) that is based on such devices (note the addition of a source resistor to provide gain compression). A variable frequency crystal oscillator (VXO) can be constructed by replacing the inductance with a quartz crystal, but the achievable frequency variation is often very small.

5.5   Negative resistance approach to oscillators

When a tuned circuit is excited by a pulse, it will ring at the resonant frequency. The circuit resistance will, however, cause a rapid damping of these oscillations. This can be overcome by introducing a device that has negative resistance in order to cancel out the circuit resistance. An example of such a device is the Gunn diode shown in Figure 5.19. A Gunn diode exhibits higher energy states for which the current carriers have lower mobility and this will cause negative resistance under suitable bias conditions (see Figure 5.20). The negative resistance can be used to cancel out the damping resistance of a tuned circuit and hence create an oscillator (see Figure 5.20 and Collin).

We can also generate negative resistance using an FET (or a BJT) and this can provide an alternative way of analysing oscillators. The circuit shown in Figure 5.21 is capable

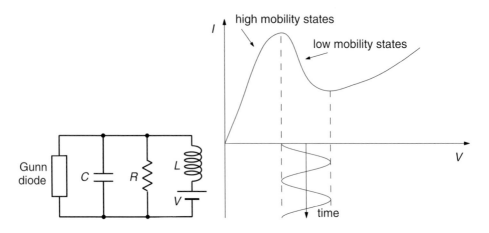

Figure 5.20   Oscillator based on the Gunn diode.

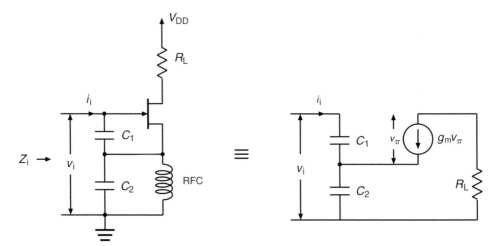

Figure 5.21   Negative resistance circuit based on an FET.

of generating negative resistance (gate bias not shown). This can be analysed through the model that is also shown in the Figure 5.21 and from which

$$v_i = \frac{i_i}{j\omega C_1} + \left(i_i + g_m \frac{i_i}{j\omega C_1} \right) \frac{1}{j\omega C_2}. \tag{5.25}$$

The input impedance Z_i will be given by

$$Z_i = \frac{v_i}{i_i} = \frac{1}{j\omega C_1} + \left(1 + \frac{g_m}{j\omega C_1} \right) \frac{1}{j\omega C_2} \tag{5.26}$$

which has a negative real part

$$R_i = \frac{-g_m}{C_1 C_2 \omega^2}. \tag{5.27}$$

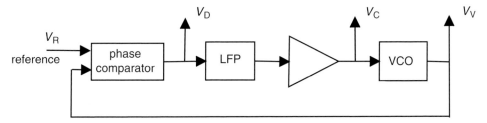

Figure 5.22 General phase locked loop.

By connecting the above impedance in parallel with an inductor, a Colpitts oscillator is formed. The magnitude of R_i will need to be larger than the intrinsic resistance of the inductor in order for oscillation to occur and this will lead to the same condition as was derived in Section 5.2.

5.6   Phase locked loops

A phase locked loop (PLL) is a feedback system in which the feedback is based on phase difference alone. These systems have a large variety of applications including frequency control and demodulation. Figure 5.22 shows a typical PLL architecture. The PLL compares the output of a voltage controlled oscillator (VCO) with a reference signal and produces a control voltage that is proportional to the phase difference between them. This voltage then adjusts the VCO such that it moves closer to the reference signal in terms of phase. When the system settles down, the VCO is basically *locked* onto the reference signal. The low-pass filter (LPF) helps remove unwanted high frequency components that are present at the phase comparator output and the amplifier ensures an adequate level of control voltage. In essence, a PLL produces a less noisy version of the reference signal, but slightly out of phase (the phase difference can be reduced by increasing the amplifier gain). The low-pass filter characteristics will be dictated by the application and, in the case of demodulation, will need to exhibit a bandwidth that is at least that of the baseband signal.

5.7   Analysis of a phase locked loop

If the reference signal v_R has the form $V_R \cos[\omega_0 t + \phi_R(t)]$ and the VCO signal v_V has the form $V_V \cos[\omega_0 t + \phi_V(t)]$, the output of the phase detector v_D will be

$$v_D(t) = k_D[\phi_R(t) - \phi_V(t)], \tag{5.28}$$

where k_D depends on the nature of the phase detector. We assume that the phase detector is linear and that ω_0 is the free running frequency of the VCO (the frequency for which $v_C = 0$). After passage through the filter (transfer function H) and the amplifier (gain A), there will result a VCO control voltage v_C with Laplace transform

$$v_C(s) = AH(s)v_D(s) = AH(s)k_D[\phi_R(s) - \phi_V(s)] \tag{5.29}$$

(argument s is the transform variable). The control voltage v_C causes a frequency shift of $\Delta\omega = k_C v_C$ and, since $\Delta\omega = d\phi_V/dt$ ($s\phi_V(s)$ in the transform plane),

$$s\phi_V(s) = k_C v_C(s) = k_C A H(s) k_D[\phi_R(s) - \phi_V(s)] \tag{5.30}$$

from which

$$\frac{\phi_V(s)}{\phi_R(s)} = \frac{k_C k_D A H(s)}{s + k_C k_D A H(s)}. \tag{5.31}$$

It is clear that, even with $H(s) = 1$, the phase variations of the VCO output will be a filtered version of the phase variations of the reference (reinterpret s as $j\omega$). Consequently, the VCO output will be a less noisy version of the reference signal.

For a single pole low-pass filter (see Chapter 8), the transfer function is given by

$$H(s) = \frac{\omega_c}{s + \omega_c}. \tag{5.32}$$

Consequently,

$$\frac{\phi_V(s)}{\phi_R(s)} = \frac{\omega_V^2}{s^2 + 2\zeta\omega_V s + \omega_V^2}, \tag{5.33}$$

where $\omega_V = \sqrt{k_C k_D A \omega_c}$ and $\zeta = \omega_c/2\omega_V$. For a sudden change ϕ_0 in the phase of the reference signal ($\phi_R(s) = \phi_0/s$ in transform terms), the transform can be inverted to yield the VCO phase response

$$\phi_V(t) = \phi_0 - \phi_0 \exp(-\zeta\omega_V t)\left[\cos\left(\omega_V\sqrt{1-\zeta^2}t\right) + \frac{\zeta}{\sqrt{1-\zeta^2}}\sin\left(\omega_V\sqrt{1-\zeta^2}t\right)\right]. \tag{5.34}$$

Figure 5.23 illustrates this response for a variety of values of ζ and suggests that, when $\zeta \ll 1$, the system will be under damped and the phase response will exhibit strong oscillations before settling into its new state. For $\zeta \gg 1$, the system will be overdamped and take a long time to settle into its new state. It is clear that a value around $1/\sqrt{2}$ for ζ represents a good compromise between speed of response and oscillation.

After switch on, if the VCO frequency is too far from that of the reference signal, the control voltage v_C will be blocked by the loop filter and the system will not lock. If the frequencies are close enough, however, there will be a control voltage that can pull the VCO into lock. The range of reference frequencies for which this can occur is known as the *capture range*. After lock has occurred, the range of frequencies over which this lock is maintained is known as the *lock range*. It turns out that both capture and lock range will be of the order of the filter bandwidth ω_c.

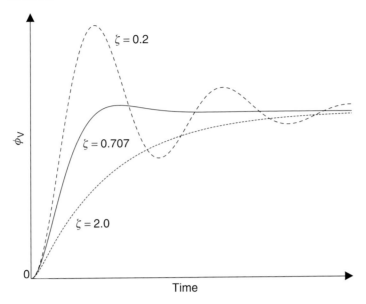

Figure 5.23   Response of a phase locked loop to a step change in reference phase.

5.8   Phase locked loop components

We have already discussed VCO circuits, but other PLL components warrant further consideration. Standard mixer circuits can be employed as phase detectors, but other options are perhaps better suited to integrated circuit implementations employing digital signals. An exclusive OR gate can provide an effective phase comparator and Figure 5.24 illustrates its operation. As the phase difference between the VCO and the reference signals increases, the filtered output voltage v_C increases until it reaches a maximum at a phase difference of π. A simple RC combination can be used to provide the loop filter, but this needs to be followed by a suitable voltage amplifier. Both the amplifier and filter functions can, however, be combined in the operational amplifier shown in Figure 5.25 (note that $v_C = v_D R_F / R_I (j\omega R_F C + 1)$).

When a phase locked loop is initially switched on, the VCO and reference frequencies can be considerably different and this can make it difficult for lock to be achieved. What is needed is an initial process in which the control voltage responds to frequency difference in order to bring the VCO sufficiently close for phase lock to occur. This is achieved using a phase/frequency detector (PFD), an example of which is shown in Figure 5.26. The circuit is based on edge-triggered D flip-flops with the normal D inputs connected to the d.c. supply and the clock inputs to the reference and VCO signals. When the non-inverted outputs of both flip-flops are high, both devices are reset. Figures 5.27a and 5.27b show the PFD output for input signals having different frequency and phase, respectively. Depending on the sign of the difference, charge is pumped in or

(a)
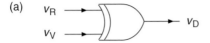

(b)
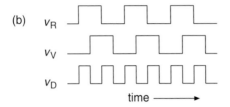

(c)
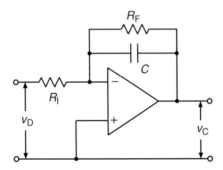

Figure 5.24   Phase comparator.

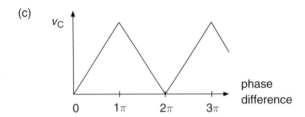

Figure 5.25   Loop filter.

out of a reservoir capacitor C (the rate is controlled by the bias on the outer transistors). This causes a voltage V that can be used to control the VCO. In a PLL system, the frequency detection mode will operate until capture is achieved and then the phase detection mode will come into play.

Figure 5.28 shows a simple PLL based on a JFET Colpitts oscillator and a dual-gate MOSFET mixer. Components R_5 and C_4 constitute a single pole low-pass filter and this can be used to tailor the characteristics of the PLL (capture and lock ranges). It should be noted, however, that the dynamics of the loop will also be affected by the level of input signal, the amplitude of oscillations, the capacitance properties of the diode and the bias of the mixer. In particular, the dependence upon input level will make it preferable to buffer the PLL input by means of a limiting amplifier.

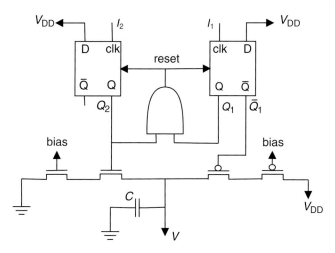

Figure 5.26 Phase/frequency detector based on D flip-flops and a charge pump.

5.9   Phase locked loop applications

Important PLL applications arise in the demodulation of RF signals. A PLL can be used to demodulate an FM signal if the PLL reference is replaced by the modulated RF signal (see Figure 5.29). Providing the oscillator frequency is a linear function of the control voltage v_C, this voltage will reproduce the original baseband signal. It is important, however, that the loop filter has a bandwidth that can accommodate this baseband signal.

An AM demodulator is a little more tricky, but can be achieved using the circuit shown in Figure 5.30. We have already seen that a product detector can be used to demodulate AM signals if the local oscillator frequency is that of the carrier. In the circuit of Figure 5.30, the PLL will lock onto the carrier of the AM signal and hence the VCO output provides a suitable local oscillator signal.

One of the most important applications of PLLs is in frequency synthesis. A frequency synthesiser is a device that can generate a large range of accurate frequencies from a single stable reference. A typical architecture is shown in Figure 5.31 and relies on the existence of accurate high speed programmable frequency dividers but, with the advent of digital techniques, such devices are readily available (Figure 5.32 shows two examples of simple frequency dividers). For a given division N, the synthesiser will lock onto the frequency Nf_r where f_r is the frequency of the reference oscillator. For FM transmission applications, it is possible to provide a modulated output by introducing a suitably scaled version of the baseband signal at point B. It should be noted, however, that the loop filter will need to be kept narrow enough for the modulation not to disturb the synthesis of the carrier frequency.

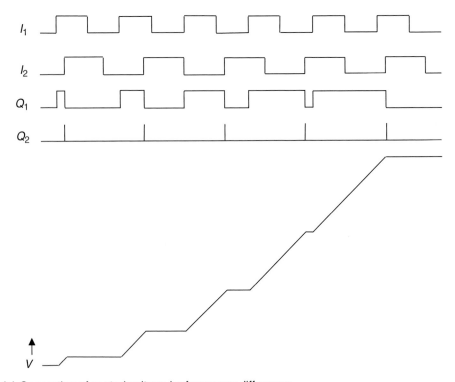

(a) Generation of control voltage by frequency difference

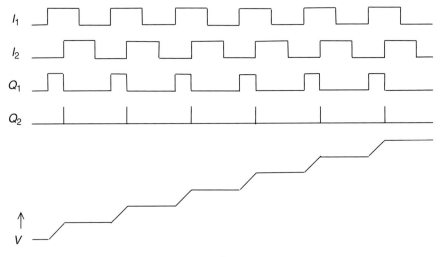

(b) Generation of control voltage by phase difference

Figure 5.27   Signal levels in the phase/frequency detector.

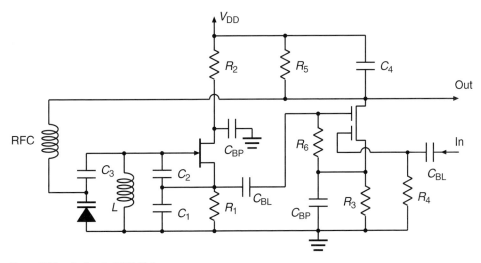

Figure 5.28   A simple FET PLL.

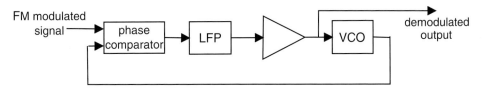

Figure 5.29   Phase locked loop FM demodulator.

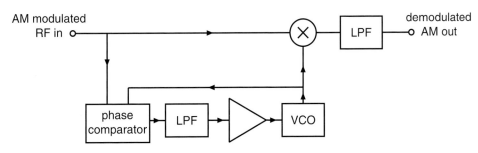

Figure 5.30   Phase locked loop AM detector.

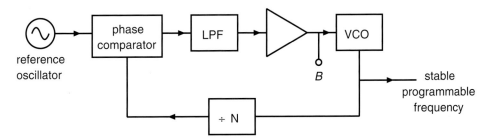

Figure 5.31   Frequency synthesiser.

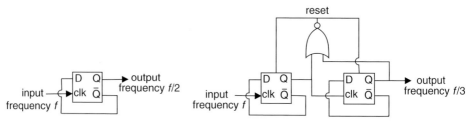

Figure 5.32   Frequency dividers based on D flip-flops.

EXERCISES

(1) Design a 15 MHz BJT Colpitts oscillator based on the common-emitter amplifier. (Assume $C_\pi = 25\,\text{pF}$, $C_\mu = 8\,\text{pF}$, $\beta = 50$ and a supply voltage of 10 V.)

(2) Design a 50 MHz JFET Colpitts oscillator based on a source follower amplifier. (Assume a 6 V supply voltage and FET parameters $V_t = -3\,\text{V}$, $K = 10^{-3}\,\text{A/V}^2$, $C_{GS} = 4\,\text{pF}$, $C_{GD} = 1.6\,\text{pF}$, $C_{DS} = 0.1\,\text{pF}$ and $r_{DS} = 30\,\text{k}\Omega$.)

(3) Draw the circuit diagram of a Hartley oscillator based on a common-source amplifier. Derive the condition for oscillation and an expression for the frequency of oscillation. (You may assume that $r_{DS} = \infty$ and $C_{GS} = C_{GD} = C_{DS} = 0$.)

(4) Design an oscillator based around a common-gate amplifier. Derive the condition for oscillation and an expression for the frequency of oscillation. (You may assume that $r_{DS} = \infty$ and $C_{GS} = C_{GD} = C_{DS} = 0$.)

(5) Show how a BJT can be configured as a negative resistance device and derive an expression for its input impedance. (You may assume that $C_\pi = C_\mu = r_x = 0$ and $r_\pi = r_o = \infty$.)

(6) For the PLL shown in Figure 5.28, choose component values such that this circuit can be used to demodulate a 2 MHz FM modulated signal that has a maximum deviation of 10 kHz (you may use two series JFETs instead of the dual-gate MOSFET). Perform a simulation of this circuit using your chosen values and a suitable circuit simulation package.

(7) Redesign the PLL shown in Figure 5.28 using an exclusive OR gate instead of the MOSFET phase detector.

(8) Redesign the PLL of Question 7 using a CMOS differential oscillator instead of the JFET oscillator.

SOURCES

Collin, R. E. 1992. *Foundations for Microwave Engineering* (2nd edn). New York: McGraw-Hill.

Everard, J. 2001. *Fundamentals of RF Circuit Design with Low Noise Oscillators*. Chichester: John Wiley.

Hayward, W. 1994. *Introduction to Radio Frequency Design*. Newark, CT: American Radio Relay League.

Lee, H. 1985. *The Design of RF CMOS Radio Frequency Integrated Circuits*. Cambridge University Press.

Ludwig, R. and Bretchko, P. 1990. *RF Circuit Design: Theory and Application*. Upper Saddle River, NJ: Prentice-Hall.

Maas, S. A. 1997. *Non-linear Microwave Circuits*. Piscataway, NJ: IEEE Press.

Pozar, D. M. 1998. *Microwave Engineering* (2nd edn). New York: John Wiley.

Pozar, D. M. 2001. *Microwave and RF Design of Wireless Systems*. New York: John Wiley.

Razavi, B. 2001. *Design of Analogue CMOS Integrated Circuits*. New York: McGraw-Hill.

Rhode, U. L. and Newkirk, D. P. 2000. *RF/Microwave Circuit Design for Wireless Applications*. New York: John Wiley.

Rhode, U. L., Whitaker, J. and Bucher, T. T. N. 1996. *Communication Receivers* (2nd edn). New York: McGraw-Hill.

Sabin, W. E. and Schoenike, E. O. (eds.). 1998. *HF Radio Systems and Circuits*. Atlanta, GA: Noble Publishing Corporation.

Sedra, A. S. and Smith, K. C. 1991. *Microelectronic Circuits* (3rd edn). Oxford University Press.

Smith, J. R. 1997. *Modern Communication Circuits*. New York: McGraw-Hill.

Straw, R. Dean (ed.). 1999. *The ARRL Handbook* (77th edn). Newark, CT: American Radio Relay League.

6 Transmission lines and scattering matrices

In propagating through free space, radio waves will suffer a reduction in amplitude as they spread outwards from the source. When a transmission must reach a large number of geographically dispersed receivers, such as in broadcast radio, there is little that can be done about this loss. If it is only required to reach one receiver, however, it is desirable to transmit all the energy to this one device. A structure for achieving this is known as a transmission line. Such structures will normally have a small uniform cross-section and can be constructed so that the loss along the line is extremely small. Transmission lines allow efficient and unobtrusive transmission over long distances. Two of the most common varieties of transmission line are the coaxial cable and the twin parallel wire. This chapter considers a simple lumped circuit model of such transmission lines and describes some important applications of these structures. The study of transmission lines leads very naturally to the concept of the reflection coefficient, a concept that provides an alternative description of impedances. Reflection coefficients generalise to the concept of scattering matrices which themselves provide an alternative means of describing multiport networks. The present chapter introduces the basic idea of the scattering matrix and shows how it can be applied to the design of small signal amplifiers at high frequencies.

6.1 The transmission line model

Figure 6.1 shows the construction of two important transmission lines, the coaxial cable and twin parallel wire. It is evident that both of these lines can be considered as a distribution of capacitance and inductance. This can be modelled, in a discrete component sense, as a set of series inductances and shunt capacitances (as illustrated in Figure 6.2). Let L and C be the inductance and capacitance per unit length. For a coaxial line

$$L = \frac{\mu}{2\pi} \ln\left(\frac{b}{a}\right) \quad \text{and} \quad C = \frac{2\pi\epsilon}{\ln\left(\frac{b}{a}\right)} \tag{6.1}$$

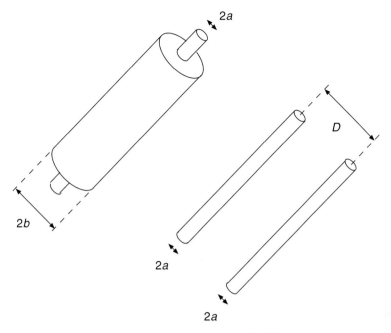

Figure 6.1   Coaxial and twin parallel wire transmission lines.

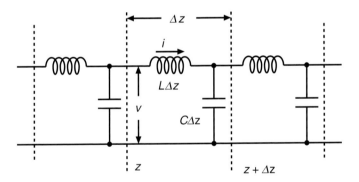

Figure 6.2   Lumped component model of transmission line.

and for a parallel wire line

$$L = \frac{\mu}{\pi} \cosh^{-1}\left(\frac{D}{2a}\right) \quad \text{and} \quad C = \frac{\pi \epsilon}{\cosh^{-1}\left(\frac{D}{2a}\right)}, \tag{6.2}$$

where ϵ and μ refer to the permittivity and permeability, respectively, of the medium between the conductors. For the circuit in Figure 6.2, Kirchhoff's voltage law implies

$$v(z, t) - L\Delta z \, \frac{\partial i(z, t)}{\partial t} - v(z + \Delta z, t) = 0 \tag{6.3}$$

and Kirchhoff's current law implies

$$i(z, t) - C\Delta z \frac{\partial v(z + \Delta z, t)}{\partial t} - i(z + \Delta z, t) = 0. \tag{6.4}$$

In the limit, as $\Delta z \to 0$,

$$\frac{\partial v(z, t)}{\partial z} = -L\frac{\partial i(z, t)}{\partial t} \tag{6.5}$$

and

$$\frac{\partial i(z, t)}{\partial z} = -C\frac{\partial v(z, t)}{\partial t}. \tag{6.6}$$

Eliminating between Equations 6.5 and 6.6, we obtain the wave equations

$$\frac{1}{c^2}\frac{\partial^2 v}{\partial t^2} - \frac{\partial^2 v}{\partial z^2} = 0 \tag{6.7}$$

and

$$\frac{1}{c^2}\frac{\partial^2 i}{\partial t^2} - \frac{\partial^2 i}{\partial z^2} = 0, \tag{6.8}$$

where $c = 1/\sqrt{CL}$. It is clear that voltage and current will propagate as waves along the transmission lines (propagation speed c). The voltage and current will behave as combinations of waves travelling in the negative and positive z directions, $v_-(z + ct)$ and $v_+(z - ct)$ for voltage and $i_-(z + ct)$ and $i_+(z - ct)$ for current. It should be noted, however, that the voltages and currents for both left and right travelling waves will be related through Equations 6.5 and 6.6. Noting that i_+ is a function of $u^+ = z - ct$ alone and i_- of $u^- = z + ct$, Equation 6.5 yields

$$\frac{\partial v_\pm}{\partial z} = \pm cL \frac{\partial i_\pm}{\partial z}. \tag{6.9}$$

We can then integrate with respect to z to obtain

$$v_\pm = \pm Z_0 i_\pm, \tag{6.10}$$

where $Z_0 = \sqrt{L/C}$ is known as the *characteristic impedance* of the transmission line. This is the fundamental relationship between current and voltage on a transmission line.

Consider a transmission line of length l that is terminated by a resistance R (Figure 6.3). If a rectangular pulse with voltage V^+ is launched into the line, after time l/c it will reach the resistor. Some of the power will be absorbed by the resistor and the remainder reflected back down the line. The reflected pulse will have voltage $V^- = \Gamma V^+$, where Γ is known as the *reflection coefficient*. Denote the ingoing and outgoing current magnitudes by I^+ and I^-, respectively. At the resistor, Ohm's law provides the relation $V = IR$ between the total current $I = I^+ + I^-$ and voltage

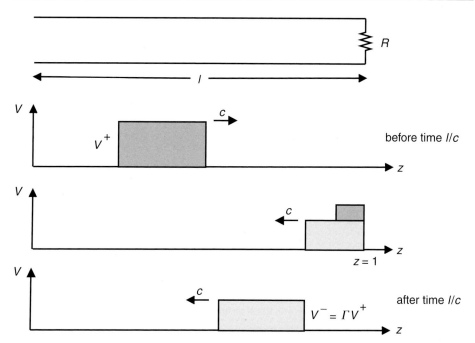

Figure 6.3 Propagation on a transmission line with resistive load.

$V = V^+ + V^-$. From Equation 6.10, $I = (V^+ - V^-)/Z_0$ and hence

$$V^+ + V^- = \frac{R}{Z_0}(V^+ - V^-). \tag{6.11}$$

This can be rearranged to yield

$$V^- = \frac{R - Z_0}{R + Z_0} V^+ \tag{6.12}$$

from which

$$\Gamma = \frac{R - Z_0}{R + Z_0}. \tag{6.13}$$

6.2   Time-harmonic variations

Since we are normally interested in signals at (or around) a particular frequency ω, an important special case is that of the time harmonic variations

$$v = \Re\{V \exp(j\omega t)\} \quad \text{and} \quad i = \Re\{I \exp(j\omega t)\}, \tag{6.14}$$

where V and I are independent of time. (It should be noted that, by means of Fourier transform techniques, it is possible to relate this special case to quite general time behaviour.) For time harmonic variations, Equation 4.5 and 4.6 imply

$$\frac{dV}{dz} = -j\omega L I \quad \text{and} \quad \frac{dI}{dz} = -j\omega C V \tag{6.15}$$

and, from these relations,

$$\frac{d^2 V}{dz^2} + \omega^2 L C V = 0 \quad \text{and} \quad \frac{d^2 I}{dz^2} + \omega^2 L C I = 0. \tag{6.16}$$

Equations 6.16 imply that, for time harmonic signals, the voltage and current amplitudes vary along the transmission line according to

$$V(z) = V_0^+ \exp(-j\beta z) + V_0^- \exp(+j\beta z) = V^+(z) + V^-(z), \tag{6.17}$$

and

$$I(z) = I_0^+ \exp(-j\beta z) + I_0^- \exp(+j\beta z) = I^+(z) + I^-(z), \tag{6.18}$$

where $\exp(-j\beta z)$ represents a harmonic wave travelling in the positive z direction and $\exp(+j\beta z)$ represents a harmonic wave travelling in the negative z direction (propagation constant β has the value $\omega\sqrt{LC}$). From Equations 6.15 we will have that $I_0^\pm = \pm V_0^\pm / Z_0$ and hence

$$I(z) = \frac{1}{Z_0}(V^+(z) - V^-(z)). \tag{6.19}$$

Now consider a loaded transmission line of length l, as illustrated in Figure 6.4. If right travelling harmonic waves (voltage amplitude V^+) enter the transmission line at $z = 0$, there can also be leftward travelling waves (voltage amplitude V^-) at this point due to reflections at the load. If V_L and I_L are the total voltage and current at the load, the load impedance Z_L will be given by

$$Z_L = \frac{V_L}{I_L}$$

$$= \frac{V_L^+ + V_L^-}{V_L^+ - V_L^-} Z_0 \tag{6.20}$$

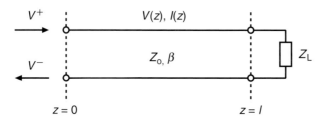

Figure 6.4   Terminated transmission line.

from which

$$V_L^- = \frac{Z_L - Z_0}{Z_L + Z_0} V_L^+, \tag{6.21}$$

where $V_L^\pm$ are the values of $V^\pm$ at $z = l$. From Equation 6.21, it can be seen that the reflection coefficient Γ_L at a general load Z_L is given by

$$\Gamma_L = \frac{Z_L - Z_0}{Z_L + Z_0}. \tag{6.22}$$

Power will not be reflected back when $\Gamma_L = 0$, that is when $Z_L = Z_0$. Under these conditions, the load is said to be matched to the line.

By means of the reflection coefficient, we can express the reflected waves in terms of the incoming waves and hence

$$I(z) = \frac{V_L^+}{Z_0} \{\exp[-j\beta(z - l)] - \Gamma_L \exp[+j\beta(z - l)]\} \tag{6.23}$$

and

$$V(z) = V_L^+ \{\exp[-j\beta(z - l)] + \Gamma_L \exp[+j\beta(z - l)]\}. \tag{6.24}$$

It will be seen that the average power at any point along the transmission line has the constant value

$$P_{av} = \frac{1}{2} \Re\{V(z)I^*(z)\} = \frac{1}{2} \frac{|V_L^+|^2}{Z_0} (1 - |\Gamma_L|^2) \tag{6.25}$$

and this is maximum when the load is matched. Furthermore, the amplitude of voltage on the line varies according to

$$|V(z)| = |V_L^+||1 + \Gamma_L \exp[2j\beta(z - l)]|, \tag{6.26}$$

that is, the voltage amplitude fluctuates between values $V_{max} = |V_L^+|(1 + |\Gamma_L|)$ and $V_{min} = |V_L^+|(1 - |\Gamma_L|)$. The *voltage standing wave ratio* (VSWR) is defined to be the ratio of the maximum and minimum voltages

$$\text{VSWR} = \frac{V_{max}}{V_{min}} = \frac{1 + |\Gamma_L|}{1 - |\Gamma_L|} \tag{6.27}$$

and is a common measure of load *mismatch* (VSWR $= 1$ for a perfect match).

An important quantity is the impedance Z_{in} that a source will see when it feeds a loaded transmission line. At the load, the total voltage and current will be given by $V_L = V_L^+ + V_L^-$ and $I_L = I_L^+ + I_L^-$ respectively, from which $V_L = V_L^+(1 + \Gamma_L)$ and $I_L = I_L^+(1 - \Gamma_L)$. Then, on noting the relationship $\exp j\theta = \cos\theta + j\sin\theta$, Equation 6.24 implies

$$V(z) = V_L \cos\beta(z - l) - jI_L Z_0 \sin\beta(z - l) \tag{6.28}$$

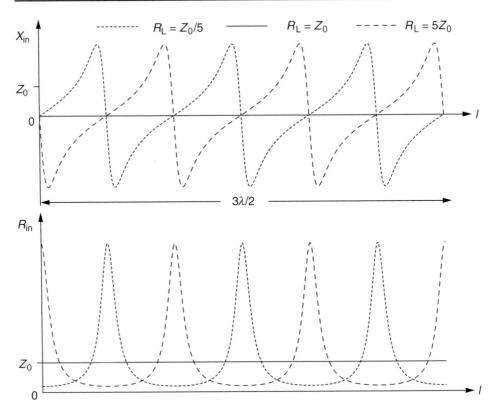

Figure 6.5   The input impedance of a loaded line ($Z_\text{in} = R_\text{in} + X_\text{in}$).

and Equation 6.23 implies

$$I(z) = I_\text{L} \cos \beta(z - l) - \text{j} \frac{V_\text{L}}{Z_0} \sin \beta(z - l). \qquad (6.29)$$

Consequently, the impedance looking into the line at $z = 0$, will be given by

$$Z_\text{in} = \frac{V(0)}{I(0)} = Z_0 \frac{Z_\text{L} \cos \beta l + \text{j} Z_0 \sin \beta l}{Z_0 \cos \beta l + \text{j} Z_\text{L} \sin \beta l} \qquad (6.30)$$

on noting that $Z_\text{L} = V_\text{L}/I_\text{L}$. Except when its value is that of the characteristic impedance Z_0, a terminating impedance Z_L will be transformed to a different value by the addition of a length of transmission line (Figure 6.5 shows some examples with totally resistive loads R_L). This result has extremely important practical consequences for RF circuits. It should also be noted that Γ_in, the reflection coefficient looking into the transmission line, will be different from the coefficient Γ_L when looking into the load itself. The relationship between these coefficients is given by

$$\Gamma_\text{in} = \Gamma_\text{L} \exp(-2\text{j}\beta l) \qquad (6.31)$$

from which it will be noted that the magnitude of the reflection coefficient remains the same and only the phase changes. This relationship is so simple that it is often easier to work in terms of reflection coefficients when considering the effects of transmission lines (reflection coefficients and impedances provide equivalent descriptions of a load).

Example A reactive load jZ_0 is connected to one end of a quarter wavelength coaxial cable with characteristic impedance Z_0. Calculate the reflection coefficient looking directly into the load and then the reflection coefficient looking into the open end of the cable. Use this last reflection coefficient to derive the impedance looking into the cable and then verify your result by directly transforming the impedance jZ_0 down the cable.

Since $Z_L = jZ_0$, we will have

$$\Gamma_L = \frac{Z_L - Z_0}{Z_L + Z_0} = \frac{jZ_0 - Z_0}{jZ_0 + Z_0} = \frac{j-1}{j+1}. \tag{6.32}$$

Transforming this reflection coefficient down the cable ($l = \lambda/4$), we have the impedance looking into the open end given by

$$\Gamma_{in} = \Gamma_L \exp\left(-2j\beta \frac{\lambda}{4}\right)$$

$$= \Gamma_L \exp\left(-2j\frac{2\pi}{\lambda}\frac{\lambda}{4}\right) = -\Gamma_L = -\frac{j-1}{j+1} \tag{6.33}$$

on noting that $\beta = 2\pi/\lambda$. Consequently,

$$Z_{in} = Z_0 \frac{1 + \Gamma_{in}}{1 - \Gamma_{in}} = Z_0 \frac{1 - \frac{j-1}{j+1}}{1 + \frac{j-1}{j+1}}$$

$$= Z_0 \frac{j + 1 - j + 1}{j + 1 + j - 1} = -jZ_0. \tag{6.34}$$

Impedance Z_L can be directly transformed down the line to impedance Z_{in} by means of expression 6.30 and, since $l = \lambda/4$,

$$Z_{in} = Z_0 \frac{jZ_0}{jZ_L} = Z_0 \frac{jZ_0}{jjZ_0} = -jZ_0 \tag{6.35}$$

as expected.

6.3   Real transmission lines

In order to carry out transmission line calculations, we will need the propagation constant β and the line impedance Z_0. For a lossless transmission line,

$$\beta = \omega\sqrt{LC} \quad \text{and} \quad Z_0 = \sqrt{\frac{L}{C}}, \tag{6.36}$$

where L and C are the inductance and capacitance per unit length. For harmonic waves, the wavelength λ on the transmission will be related to β through $\beta = 2\pi/\lambda$ and it should also be noted that wavelength λ is related to the wavelength λ_0 of electromagnetic waves in free space through

$$\lambda = \frac{1}{\sqrt{\epsilon_{\text{eff}}}}\lambda_0, \tag{6.37}$$

where ϵ_{eff} is the *effective relative permittivity* of the transmission line (note that $c = c_0/\sqrt{\epsilon_{\text{eff}}}$, where $c_0 = 3 \times 10^8$ m/s is the propagation speed in free space). For a *coaxial* line, (as in Figure 6.1)

$$Z_0 = \frac{1}{2\pi}\sqrt{\frac{\mu}{\epsilon}}\ln\frac{b}{a} \tag{6.38}$$

and

$$\epsilon_{\text{eff}} = \frac{\mu\epsilon}{\mu_0\epsilon_0}, \tag{6.39}$$

where μ and ϵ are the permeability, and permittivity respectively, of the region between the conductors.

Another important transmission line is the *microstrip*, an example of which is shown in Figure 6.6. This variety can be fabricated using printed circuit techniques and is extremely useful for distributing power around high frequency printed circuit boards. Assuming that $w/h > 1$, it can be shown (see Collin) that

$$Z_0 \approx \frac{120\pi}{\sqrt{\epsilon_{\text{eff}}}\left[1.393 + \frac{w}{h} + \frac{2}{3}\ln\left(\frac{w}{h} + 1.444\right)\right]}, \tag{6.40}$$

where $\epsilon_{\text{eff}} = (\epsilon_r + 1)/2 + [(\epsilon_r - 1)/2](1 + 12h/w)^{-1/2}$ and ϵ_r is the relative permittivity of the substrate. For the purposes of design, however, we are normally given h, ϵ_r and

substrate

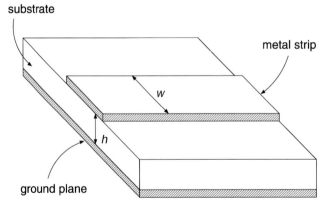

Figure 6.6 Microstrip transmission line.

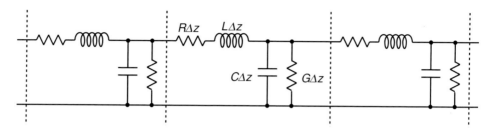

Figure 6.7   Realistic model of transmission line.

Z_0. The appropriate strip width w can then be found (Wheeler) by means of expression

$$\frac{w}{h} \approx \frac{8\left(\frac{7\epsilon_r+4}{11\epsilon_r}A + \frac{\epsilon_r+1}{0.81\epsilon_r}\right)^{1/2}}{A}, \tag{6.41}$$

where $A = \exp[(Z_0/42.4)\sqrt{\epsilon_r + 1}] - 1$.

Thus far our considerations have concentrated on transmission lines in which there is no loss. Practical transmission lines will, however, exhibit losses and the model shown in Figure 6.7 will be more appropriate. In this figure, G is the conductance per unit length resulting from losses in the material separating the transmission line conductors and R is the resistance per unit length due to the finite conductivity of those conductors. For a microstrip transmission line, $R = (2/w)\sqrt{\omega\mu_0/2\sigma}$, where σ is the conductivity of the metal conductors and $G = \omega C \tan \delta$, where $\tan \delta$ is the *loss tangent* of the substrate (a parameter that is normally supplied by the printed circuit board manufacturer). The propagation constant and characteristic impedances will be given by

$$\beta = \omega\sqrt{LC}\sqrt{\left(1 + \frac{R}{j\omega L}\right)\left(1 + \frac{G}{j\omega C}\right)}$$

and

$$Z_0 = \sqrt{\frac{R + j\omega L}{G + j\omega C}}$$

from which it will be noted that both quantities are now complex. Since most practical transmission lines will be operated such that $R \ll \omega L$ and $G \ll \omega C$, the approximations $\beta \approx \omega\sqrt{LC}[1 - (j/2)(R/\omega L + G/\omega C)]$ and $Z_0 \approx \sqrt{L/C}$ are usually valid. For a right travelling harmonic wave on the transmission line

$$v(z, t) = \exp(\Im\{\beta\}z)\,\Re\{V^+ \exp j(\omega t - \Re\{\beta\}z)\}, \tag{6.42}$$

and

$$i(z, t) = \exp(\Im\{\beta\}z)\,\Re\left\{\frac{V^+}{Z_0} \exp j(\omega t - \Re\{\beta\}z)\right\}, \tag{6.43}$$

where V^+ is the amplitude of v at $z = 0$. It will be noted that the imaginary part of β causes a loss of amplitude that accumulates over distance and, in travelling a distance l along a transmission line, power will be attenuated by an amount $8.686l|\Im\{\beta\}|$ in terms of decibels.

Example A coaxial cable is advertised as having an impedance of 50 Ω, a capacitance of 100 pF/m and a loss of 10 dB/100 m at 500 MHz. Calculate β, c and L for this cable at a frequency of 500 MHz.

Since $Z_0 = \sqrt{L/C}$, we will have $L = Z_0^2 C = 50^2 \times 10^{-10} = 0.25\,\mu$ H/m. The propagation speed on the cable will be given by $c = 1/\sqrt{LC} = 2 \times 10^8$ m/s. The real part of the propagation constant is given by $\Re\{\beta\} = 2\pi f\sqrt{LC} = 5\pi$. The complex part of β can be obtained from the fact that the attenuation per unit length is given by $8.686|\Im\{\beta\}|$ and, since this must have the value 0.1 dB/m, $\Im\{\beta\} = -0.0115$. Consequently, $\beta = 5\pi - 0.0115j$.

6.4   Impedance transformation

We have already noted (Equation 6.30) that an impedance Z_L will be transformed down a transmission line of length l according to

$$Z_I = \frac{Z_L + jZ_0 \tan \beta l}{Z_0 + jZ_L \tan \beta l} Z_0, \tag{6.44}$$

where Z_0 is the characteristic impedance of the line. Three important special cases are:

The short-circuited line. (Figure 6.8).
This exhibits impedance $Z_I = jZ_0 \tan \beta l$ and, as a consequence, $Z_I = \infty$ for $l = \lambda/4$.
The open-circuited line. (Figure 6.9).
This exhibits impedance $Z_I = -jZ_0 \cot \beta l$ and, as a consequence, $Z_I = 0$ for $l = \lambda/4$.
The $\lambda/4$ transformer. (Figure 6.10).
This exhibits impedance $Z_I = Z_0^2/Z_L$ and, as a consequence, a value of $Z_0 = \sqrt{Z_L Z_I}$ will achieve a transformation of Z_L to Z_I.

The above results are important in that they indicate how we might use transmission lines to fabricate impedances and also to transform impedances. In particular,

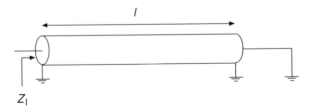

Figure 6.8 The short-circuited line.

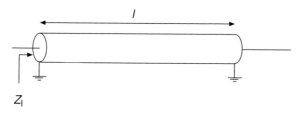

Figure 6.9   The open-circuited line.

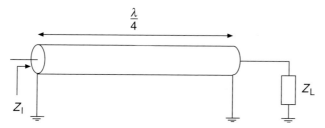

Figure 6.10   The $\lambda/4$ transformer.

by means of microstrip lines, we can fabricate components for printed circuit board implementations of high frequency designs. An example is the fabrication of an RF choke for isolating the d.c. supply to an RF transistor. The supply line can be made out of a microstrip that is RF shorted to ground (by means of a capacitor) at a quarter wavelength from the transistor. This causes the transistor to see an infinite impedance at the desired RF frequency. (This technique is used in the RF amplifier example at the end of this chapter.) It will also be noted that, around the frequency for which a transmission line is a quarter wavelength, open- and short-circuited lengths of line will behave as series and parallel resonant LC circuits, respectively. Transmission line resonators can be constructed with particularly high values of Q ($Q \approx \omega_r L/R$, where ω_r is the resonant frequency, L the inductance per unit length and R the resistance per unit length). In particular, Q values of several thousand can be achieved at gigahertz frequencies where combinations of inductors and capacitors will cease to be effective.

It is clear from Equation 6.44 that we can generate any impedance we like from a suitably terminated length of transmission line. As a consequence, by analogy with L-networks, it should be possible to fabricate matching networks by means of short lengths of transmission line (*stubs*). Using the network shown in Figure 6.11, we can transform the impedance Z_L into impedance Z_I. This is achieved as follows:

1. Choose length l such that the load admittance Y_L ($= Z_L^{-1}$) is transformed to an admittance with correct conductance (i.e., $\Re\{Z_I^{-1}\}$).
2. Choose length s such that the stub has susceptance that supplements that of transformed admittance to produce the required susceptance (i.e., $\Im\{Z_I\}^{-1}$).

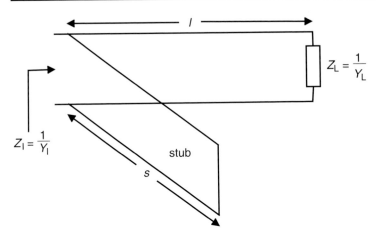

Figure 6.11   Stub matching.

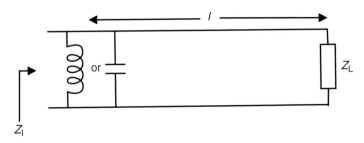

Figure 6.12   Replacing the stub by a lumped component.

Note that it is sometimes necessary to use an open-circuit stub (rather than the shorted stub of Figure 6.11) to get the correct susceptance when there is a risk that the shorted stub will become too long. If the stub cannot be brought down to a manageable length, an alternative is to use a lumped inductance, or capacitance, to achieve the same effect as the stub (see Figure 6.12).

Example   Match a load with impedance $Z_L = 15 + j10 \, \Omega$ to a transmission line with characteristic impedance $Z_0 = 50 \, \Omega$. (See Figure 6.13.)

Choose all transmission lines to have impedance $Z_0 = 50 \, \Omega$ and note that $Y_1 = 0.02$ S. Then, at distance x from Z_L,

$$Z(x) = Z_0 \frac{Z_L + jZ_0 \tan \beta x}{Z_0 + jZ_L \tan \beta x} = \frac{1}{Y(x)}. \tag{6.45}$$

The table below shows how the impedance of the source is transformed along the transmission line (x is the distance from the source in transmission line wavelengths).

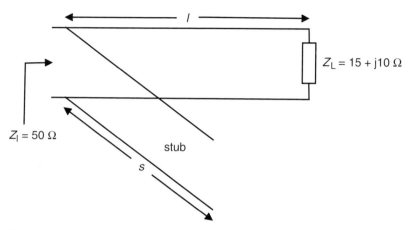

Figure 6.13 Matching example.

x	Z	Y
0.00λ	15.0 + j10.0	0.046 − j0.031
0.01λ	15.4 + j13.0	0.038 − j0.032
0.02λ	16.0 + j16.1	0.031 − j0.031
0.03λ	16.7 + j19.3	0.026 − j0.030
0.04λ	17.7 + j22.6	0.021 − j0.027
0.05λ	18.8 + j26.1	0.018 − j0.025

When $x = 0.044\lambda$, $\Re(Y) = 0.020$ and so $l = 0.044\lambda$. We will now need to add an open stub that has susceptance $B_{stub} = 0.0263$ in order to cancel out the reactive part of Z. This implies a reactance of $X_{stub} = -1/B_{stub} = -38\ \Omega$ for the stub where it joins onto the main line. The length s of stub that will yield this reactance is derived from

$$X_{stub} = -Z_0 \cot \beta s \tag{6.46}$$

and will be 0.147λ for the above value of X_{stub}. (Note that the lengths are all in terms of transmission line wavelengths and we will need the operating frequency, and information about the transmission line, in order to calculate geometric lengths.)

6.5   Reflection coefficients

For a one-port network, the reflection coefficient Γ provides an alternative description of the network through the relationship between ingoing V^+ and outgoing V^- waves (see Figure 6.14)

$$V^- = \Gamma V^+. \tag{6.47}$$

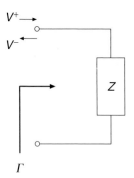

Figure 6.14   Reflection coefficient description of a one-port network.

This contrasts with the impedance description

$$V = ZI \tag{6.48}$$

which provides a relationship between the total voltage V and current I at the network terminals. These two descriptions are related through

$$\Gamma = \frac{Z - Z_0}{Z + Z_0} \tag{6.49}$$

and

$$Z = Z_0 \frac{1 + \Gamma}{1 - \Gamma}, \tag{6.50}$$

where Z_0 is the characteristic impedance of the transmission line that connects to the network. (Note that the reflection coefficient Γ is always defined with respect to a characteristic impedance Z_0, usually $50\,\Omega$.) A popular means of representing reflection coefficients is as points on a Smith chart (vertical and horizontal components represent the imaginary and real parts of Γ, respectively). The chart can be used to represent the variations of Γ with respect to a parameter (frequency, for example) in the form of a curve plotted on the chart. Figure 6.15 shows a complex tuned circuit and its associated Smith chart. It will be noted that there is a series resonance at 7.16 MHz and a parallel resonance at 10 MHz. Smith charts often have curves of constant resistance (R) and reactance (X) superimposed (see Figure 6.16) to allow for easy conversion between Γ and associated impedance Z (note that the reactance and resistance have been scaled on the characteristic impedance Z_0).

It is instructive to consider source and load matching in terms of reflection coefficients. Consider a source of impedance Z_S that is connected to a load of impedance Z_L by a length l of transmission line (see Figure 6.17). We will denote the source and load reflection coefficients by Γ_S and Γ_L, respectively and the transmission line characteristic impedance by Z_0. If the source has open-circuit voltage V_S, there will be a voltage $V = Z_{in} V_S / (Z_{in} + Z_S)$ at the input to the transmission line ($Z_{in} = Z_0 (Z_L + jZ_0 \tan \beta l)/(Z_0 + jZ_L \tan \beta l)$ is the impedance looking into the line).

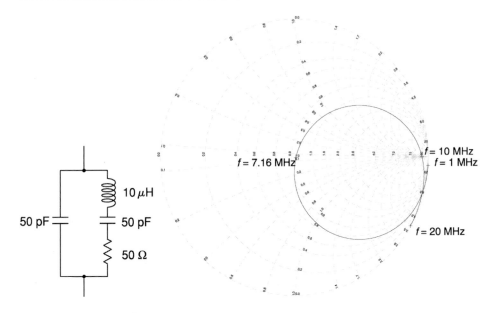

Figure 6.15   A complex tuned circuit and associated Smith chart.

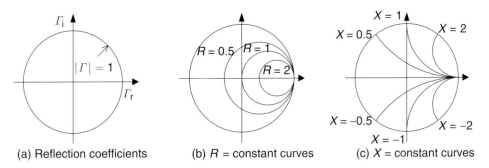

(a) Reflection coefficients   (b) R = constant curves   (c) X = constant curves

Figure 6.16   Smith charts.

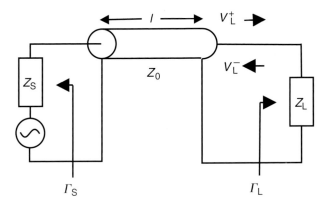

Figure 6.17   Source and load connected by a transmission line.

A voltage wave of amplitude $V^+ = V/(1 + \Gamma_{\text{in}})$ will enter the transmission line ($\Gamma_{\text{in}} = \Gamma_L \exp(-2j\beta l)$) and this will cause a wave of amplitude $V_i^+ = V^+ \exp(-j\beta l)$ to enter the load. If we turn on the source, the ingoing voltage wave amplitude V_L^+ at the load will have initial value V_i^+ and the outgoing wave an initial amplitude of $V_L^- = \Gamma_L V_i^+$. The outgoing wave will, in turn, be reflected at the source and return to the load with amplitude $\Gamma_L \Gamma_S V_i^+ \exp(-2j\beta l)$. Consequently, the ingoing voltage V_L^+ will now have amplitude $V_i^+ (1 + \Gamma_L \Gamma_S \exp(-2j\beta l))$. These reflections will continue to build up until the system settles into equilibrium. At equilibrium, the ingoing voltage wave at the load will have amplitude

$$V_L^+ = V_i^+ [1 + \Gamma_L \Gamma_S \exp(-2j\beta l) + \Gamma_L^2 \Gamma_S^2 \exp(-4j\beta l) + \Gamma_L^3 \Gamma_S^3 \exp(-6j\beta l) + \cdots]$$

$$= \frac{V_i^+}{1 - \Gamma_L \Gamma_S \exp(-2j\beta l)} \tag{6.51}$$

and the outgoing wave will have amplitude

$$V_L^- = \frac{\Gamma_L V_i^+}{1 - \Gamma_L \Gamma_S \exp(-2j\beta l)}. \tag{6.52}$$

In addition, the total voltage across the load will be

$$V_L = V_L^+ + V_L^- = V_i^+ \frac{1 + \Gamma_L}{1 - \Gamma_L \Gamma_S \exp(-2j\beta l)} \tag{6.53}$$

and the total current flowing into the load

$$I_L = \frac{V_L^+ - V_L^-}{Z_0} = \frac{V_i^+}{Z_0} \frac{1 - \Gamma_L}{1 - \Gamma_L \Gamma_S \exp(-2j\beta l)}. \tag{6.54}$$

The power dissipated in the load will therefore be given by

$$P_L = \frac{1}{2} \Re\{I_L^* V_L\} = \frac{|V_i^+|^2}{2Z_0} \frac{1 - \Gamma_L^* \Gamma_L}{[(1 - \Gamma_L \Gamma_S \exp(-2j\beta l)][1 - \Gamma_L^* \Gamma_S^{z*} \exp(2j\beta l)]}. \tag{6.55}$$

Dissipation will be maximum when $\partial P_L/\partial \Gamma_L = 0$ and this implies that $\Gamma_S^* = \Gamma_L \exp(-2j\beta l)$. Since $\Gamma_{\text{in}} = \Gamma_L \exp(-2j\beta l)$ is the reflection coefficient at the load (Γ_L) when transformed down the transmission line, we have the conjugate matching condition $\Gamma_S^* = \Gamma_{\text{in}}$. From the relation between reflection coefficients and load impedance, this will imply the impedance form of the conjugate matching condition $Z_S^* = Z_{\text{in}}$.

6.6   S parameters

The concept of reflection coefficients can be generalised to two-port networks (and higher) by means of the scattering matrix (sometimes known as the S matrix). For a two-port network (Figure 6.18) that is excited by ingoing voltage waves of magnitude

Figure 6.18   *S* matrix description of two-port network.

Figure 6.19   *Z* matrix description of two-port network.

V_1^+ and V_2^+, the outgoing waves V_1^- and V_2^- can be found from the *S* matrix relation

$$V^- = SV^+ \tag{6.56}$$

where

$$V^\pm = \begin{pmatrix} V_1^\pm \\ V_2^\pm \end{pmatrix}. \tag{6.57}$$

This is an alternative to the *Z* matrix, a network description that provides a relationship between the total voltage and current at the network ports. For a two-port network (see Figure 6.19),

$$V = ZI, \tag{6.58}$$

where

$$V = \begin{pmatrix} V_1 \\ V_2 \end{pmatrix} \quad \text{and} \quad I = \begin{pmatrix} I_1 \\ I_2 \end{pmatrix}. \tag{6.59}$$

The ingoing and outgoing waves are related to the total voltage and current at the network ports through $V = V^+ + V^-$ and $I = I^+ + I^-$. Since $I = (V^+ - V^-)/Z_0$ (noting relation 6.10), we will have

$$V^\pm = \frac{1}{2}(V \pm Z_0 I), \tag{6.60}$$

where Z_0 is the characteristic impedance of the transmission lines that connect to the ports. (Note that the *S* matrix is always defined with respect to a characteristic impedance Z_0, usually 50 Ω.) The *Z* and *S* matrix descriptions are related through

$$S = (Z + Z_0 I)^{-1}(Z - Z_0 I) \tag{6.61}$$

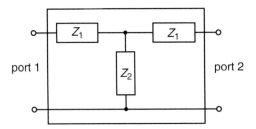

Figure 6.20 Symmetric T-network example.

and

$$Z = Z_0(I - S)^{-1}(I + S), \tag{6.62}$$

where I is the unit matrix. This is a generalisation of the relationship between the reflection coefficient of a one-port network and its impedance. In theory, the S matrix coefficients can be determined according to

$$S_{ij} = \left. \frac{V_i^-}{V_j^+} \right|_{V_k^+ = 0} \qquad \text{for all } k \neq j, \tag{6.63}$$

that is drive port j with V_j^+ and measure V_i^- coming out of port i. All ports other than j are matched ($V_k^+ = 0$ for $k \neq j$) which can be achieved by loading these ports with the characteristic impedance Z_0. We will look at the practical implementation of this procedure in Section 6.8.

Example Find the Z and S matrices corresponding to the symmetric T-network shown in Figure 6.20.

The Z matrix description provides the following relationships between the voltages and currents at the network ports

$$V_1 = Z_{11} I_1 + Z_{12} I_2$$
$$V_2 = Z_{21} I_1 + Z_{22} I_2. \tag{6.64}$$

Since the network is symmetric, we will have $Z_{12} = Z_{21}$ and $Z_{11} = Z_{22}$. If we leave port 2 open circuit ($I_2 = 0$) and measure the impedance at port 1 (i.e., V_1/I_1), we will, in fact, be measuring Z_{11}. From the network circuit, however, it is clear that this impedance will be $Z_1 + Z_2$. Consequently, $Z_{11} = Z_{22} = Z_1 + Z_2$. If we now impose voltage V_1 at port 1 and measure the open-circuit voltage V_2 at port 2, the current I_1 at port 1 will be related to this voltage by $V_2 = Z_{21} I_1$. Furthermore, by considering the network circuit, it is clear that we have a voltage divider and that V_2 will be related to V_1 through $V_2 = V_1 Z_2/(Z_1 + Z_2)$. As a consequence, $Z_{11} = V_1/I_1 = \{[(Z_1 + Z_2)V_2]/Z_2\}(Z_{21}/V_2)$. Since $Z_{11} = Z_1 + Z_2$, we will therefore have that $Z_{21} = Z_2 = Z_{12}$.

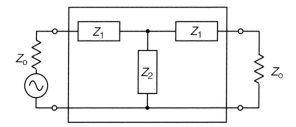

Figure 6.21   Configuration for the derivation of S parameters.

The S matrix description provides the following relationships between incoming and outgoing voltages

$$V_1^- = S_{11} V_1^+ + S_{12} V_2^+$$
$$V_2^- = S_{21} V_1^+ + S_{22} V_2^+. \tag{6.65}$$

Since the network is symmetric, we have that $S_{21} = S_{12}$ and $S_{11} = S_{22}$. We place a matched load at port 2 and a matched source at port 1 (see Figure 6.21). Due to the matched load at port 2, we will have $V_2^+ = 0$ and so, from Equation 6.65, S_{11} will be the reflection coefficient at port 1. By considering the circuit of Figure 6.21, the impedance looking into port 1 will be

$$Z_{in} = Z_1 + Z_2 \parallel (Z_1 + Z_0)$$
$$= Z_1 + \frac{Z_2(Z_1 + Z_0)}{Z_0 + Z_1 + Z_2} = \frac{Z_1^2 + Z_0 Z_1 + 2Z_1 Z_2 + Z_0 Z_2}{Z_0 + Z_1 + Z_2}. \tag{6.66}$$

Consequently, S_{11} and S_{22} will be given by

$$S_{11} = S_{22} = \Gamma_{in} = \frac{Z_{in} - Z_0}{Z_{in} + Z_0}$$
$$= \frac{Z_1^2 + 2Z_1 Z_2 - Z_0^2}{Z_1^2 + 2Z_0 Z_1 + 2Z_1 Z_2 + 2Z_0 Z_2 + Z_0^2}. \tag{6.67}$$

Since port 2 is matched we have $V_2^+ = 0$ and hence $V_2 = V_2^-$. Consequently, regarding the network of Figure 6.21 as a sequence of voltage dividers,

$$V_2^- = V_2 = V_1 \frac{Z_{in} - Z_1}{Z_{in}} \frac{Z_0}{Z_1 + Z_0}. \tag{6.68}$$

At port 1, $V_1 = V_1^+ + V_1^- = V_1^+ + S_{11} V_1^+$ and so $V_1^+ = V_1/(1 + S_{11})$. Consequently,

$$S_{21} = \frac{V_2^-}{V_1^+}$$
$$= V_1 \frac{Z_{in} - Z_1}{Z_{in}} \frac{Z_0}{Z_1 + Z_0} \frac{1 + S_{11}}{V_1}. \tag{6.69}$$

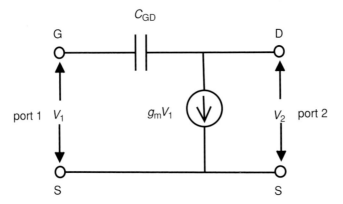

Figure 6.22   A simple two-port model of an FET.

and, after some algebra,

$$S_{12} = S_{21} = \frac{2Z_2 Z_0}{Z_1^2 + 2Z_0 Z_1 + 2Z_1 Z_2 + 2Z_0 Z_2 + Z_0^2}. \tag{6.70}$$

Example   Figure 6.22 shows a very simple two-port small signal model of an FET (suitable when the Miller effect dominates). Calculate the S matrix coefficients for this simple model.

Let $Z_{GD} = 1/j\omega C_{GD}$ be the impedance of the gate-to-drain capacitance, then the gate-to-drain current will be given by

$$I_{GD} = \frac{V_1 - V_2}{Z_{GD}}, \tag{6.71}$$

where V_1 and V_2 are the voltages at the input and output (ports 1 and 2, respectively). Let port 2 be matched (loaded by the characteristic impedance Z_0), then

$$V_2 = (I_{GD} - g_m V_1)Z_0. \tag{6.72}$$

Eliminating V_2 between Equations 6.71 and 6.72, we obtain

$$I_{GD} = \frac{1 + g_m Z_0}{Z_{GD} + Z_0} V_1 \tag{6.73}$$

and so Z_1, the impedance looking into port 1, is given by

$$Z_1 = \frac{V_1}{I_{GD}} = \frac{Z_{GD} + Z_0}{1 + g_m Z_0}. \tag{6.74}$$

Since S_{11} will be the reflection coefficient looking into port 1

$$S_{11} = \frac{Z_1 - Z_0}{Z_1 + Z_0} = \frac{Z_{GD} - g_m Z_0^2}{Z_{GD} + 2Z_0 + g_m Z_0^2}. \tag{6.75}$$

(readers should verify these for themselves), together with Equations 6.104 and 6.105, will then yield

$$G_T = \frac{1 - |\Gamma_L|^2}{|1 - S_{22}\Gamma_L|^2}|S_{21}|^2 \frac{1 - |\Gamma_S|^2}{|1 - S_{11}\Gamma_S|^2}. \qquad (6.107)$$

The first quotient represents the output network gain and the second quotient the input network gain. In the general case ($S_{12} \neq 0$),

$$G_T = \frac{|S_{21}|^2(1 - |\Gamma_S|^2)(1 - |\Gamma_L|^2)}{|(1 - S_{11}\Gamma_S)(1 - S_{22}\Gamma_L) - S_{12}S_{21}\Gamma_L\Gamma_S|^2}. \qquad (6.108)$$

As mentioned in Chapter 3, there is always the potential for an amplifier to become unstable. In terms of the reflection coefficients Γ_{in} and Γ_{out}, this will mean that $|\Gamma_{in}| > 1$ and/or $|\Gamma_{out}| > 1$. A circuit is said to be *unconditionally stable* if $|\Gamma_{in}| < 1$ for $|\Gamma_L| \leq 1$ and $|\Gamma_{out}| < 1$ for $|\Gamma_S| \leq 1$ (i.e., it is stable for all passive terminations). For the unilateral case, this will require that $|S_{11}| < 1$ and $|S_{22}| < 1$. If $S_{12} \neq 0$ there will be feedback and an increased possibility of instability. In general, if $S_{22} > 1$ and/or $S_{11} > 1$ there is no chance of unconditional stability since a matched source and load ($\Gamma_S = 0$ and $\Gamma_L = 0$) will always lead to $|\Gamma_{out}| > 1$ and/or $|\Gamma_{in}| > 1$. The values of Γ_L for which the amplifier will be unstable ($|\Gamma_{in}| > 1$) will, from Equation 6.97, be bounded by the circle $|S_{11} - \Delta\Gamma_L| = |1 - S_{22}\Gamma_L|$ on the Γ_L Smith chart. In a similar fashion, the Γ_S for which the amplifier will be unstable ($|\Gamma_{out}| > 1$) will be bounded by the circle $|S_{22} - \Delta\Gamma_S| = |1 - S_{11}\Gamma_S|$ on the Γ_S Smith chart. Figure 6.25 illustrates the regions of instability on Smith charts. The shaded regions contain the Γ_S and Γ_L for which the amplifier is stable at input and output, respectively (the figures assume that $|S_{22}| < 1$ and $|S_{11}| < 1$, respectively). We need a simple technique to check for unconditional stability. If the circuit passes the test, stability is no longer a concern of the designer. Otherwise, the load and source impedance will need to be carefully chosen such that the amplifier remains in the stable region ($|\Gamma_{out}| < 1$ and $|\Gamma_{in}| < 1$). A useful measure

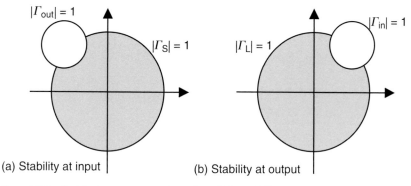

(a) Stability at input   (b) Stability at output

Figure 6.25   Smith chart representation of amplifier stability.

of stability is the Rollet factor

$$K = \frac{1 + |\Delta|^2 - |S_{11}|^2 - |S_{22}|^2}{2|S_{12}S_{21}|} \qquad (6.109)$$

and it turns out that the amplifier will be unconditionally stable if $K > 1$ and $|\Delta| < 1$.

It is not always best to design an amplifier for optimum power transfer since an increase in gain can sometimes be negated by an increase in amplifier noise. We have already seen in Chapter 3 that noise performance can be optimised by suitable matching. The noise performance is normally described through the noise factor F and this, as noted in Chapter 1, will depend on the source impedance. In terms of the source reflection coefficient, the noise factor will be given by

$$F = F_{min} + \frac{4R_N}{Z_0} \frac{|\Gamma_S - \Gamma_{opt}|^2}{(1 - |\Gamma_S|^2)|1 + \Gamma_{opt}|^2}, \qquad (6.110)$$

where Γ_{opt}, R_N and F_{min} are parameters supplied by the manufacturer for a particular transistor (Z_0 is the reference impedance for the definition of S parameters and R_N is the equivalent *noise resistance* of the circuit). It is clear that we will need $\Gamma_S = \Gamma_{opt}$ to achieve the best noise factor.

The following are three possible amplifier design strategies:

1. **Low noise and specified gain:** Γ_S is chosen for the lowest noise and then Γ_L adjusted to achieve the required gain. If stability cannot be achieved with $\Gamma_S = \Gamma_{opt}$, it might be necessary to accept a slightly less optimal value for Γ_S.
2. **Broadband with specified gain:** Choose Γ_S and Γ_L to yield the required gain at the band edges.
3. **Maximum gain:** Conjugate match at both input and output.

Example Design a maximum gain 500 MHz amplifier for 50 Ω source and load impedances using a 2N5179 BJT transistor.

A 2N5179 BJT amplifier, operating at 500 MHz with 5 mA collector current, has typical S matrix coefficients

$$\left. \begin{array}{ll} S_{11} = 0.3\angle235°, & S_{21} = 2.2\angle75° \\ S_{12} = 0.082\angle64°, & S_{22} = 0.7\angle335° \end{array} \right\} Z_0 = 50\,\Omega.$$

Consequently,

$$|\Delta| = |S_{11}S_{22} - S_{12}S_{21}| = 0.20$$

and

$$K = \frac{1 + |\Delta|^2 - |S_{11}|^2 - |S_{22}|^2}{2|S_{12}S_{21}|} = 1.71.$$

Since $|\Delta| < 1$ and $K > 1$, the amplifier will be unconditionally stable at 500 MHz.

Now consider the input reflection coefficient

$$\Gamma_{in} = S_{11} + \frac{S_{12}S_{21}\Gamma_L}{1 - S_{22}\Gamma_L} \tag{6.111}$$

and the output coefficient

$$\Gamma_{out} = S_{22} + \frac{S_{12}S_{21}\Gamma_S}{1 - S_{11}\Gamma_S}. \tag{6.112}$$

For maximum gain, we require conjugate match, that is $\Gamma_{in} = \Gamma_S^*$ and $\Gamma_{out} = \Gamma_L^*$. Equations 6.111 and 6.112 will yield a pair of non-linear simultaneous equations for the values of Γ_S and Γ_L and these can be solved numerically. Sometimes, however, the value of S_{12} is small enough for the unilateral assumption ($S_{12} = 0$) to be acceptable. In this case,

$$\Gamma_S = S_{11}^* \quad \text{and} \quad \Gamma_L = S_{22}^*. \tag{6.113}$$

Whether this assumption is valid depends upon the value of the unilateral figure of merit

$$U = \frac{|S_{12}||S_{21}||S_{22}||S_{11}|}{(1 - |S_{11}|^2)(1 - |S_{22}|^2)}. \tag{6.114}$$

$2U$ is the relative error in gain under the unilateral assumption and U should be less than 0.1 if the unilateral assumption is used. In the case of the above BJT, $U = 0.08$ and the unilateral assumption is a reasonable approximation. Consequently,

$$Z_S = \frac{1 + \Gamma_S}{1 - \Gamma_S} Z_0 \approx \frac{1 + S_{11}^*}{1 - S_{11}^*} Z_0 = 31.7 + j17.1\,\Omega$$

and

$$Z_L = \frac{1 + \Gamma_L}{1 - \Gamma_L} Z_0 \approx \frac{1 + S_{22}^*}{1 - S_{22}^*} Z_0 = 115.3 + j133.8\,\Omega.$$

Consider the microstrip realisation of the amplifier shown in Figure 6.26. Emitter (e) is connected to the ground plane on the underside of the printed circuit board (PCB), as are the three bypass capacitors. The $\lambda/4$ stubs connect the transistor to the d.c. supplies, but present an infinite impedance at the required RF frequency (i.e., the stubs act as RF chokes). The L matching networks transform the input and output impedances of the transistor ($Z_{in} = 31.7 - j17.1\,\Omega$ and $Z_{out} = 115.3 - j133.8\,\Omega$) to 50 Ω (hence achieving a conjugate match to the desired source and load). For the *output* L-network, $s_1 = 0.152\lambda$ and $s_2 = 0.075\lambda$ and, for the *input* L-network, $l_1 = 0.177\lambda$ and $l_2 = 0.089\lambda$ (note that λ is the wavelength on the transmission line).

We assume the PCB to have an FR4 substrate ($\epsilon_r = 4.4$) with a thickness of 3 mm. The tracks will need to be 5.7 mm wide in order to ensure a 50 Ω characteristic impedance (note that the effective relative permittivity ϵ_{eff} will be 3.33). Since the wavelength on the microstrip will be 33 cm (i.e., $\lambda = \lambda_0 \epsilon_{eff}^{-1/2}$ where λ_0 is the wavelength in free space),

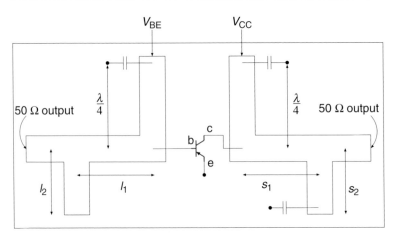

Figure 6.26 Final amplifier design.

we obtain the microstrip lengths $s_1 = 5$ cm, $s_2 = 2.5$ cm, $l_1 = 5.8$ cm and $l_2 = 2.9$ cm. (Note that the $\lambda/4$ stubs will have the length 8.25 cm.) For an open stub, there is an end effect which increases the effective geometric length (see Fooks and Zakarevicins) by

$$\Delta = 0.412 \frac{\epsilon_{\mathrm{eff}} + 0.3}{\epsilon_{\mathrm{eff}} - 0.258} \frac{w + 0.264h}{w + 0.8h} h \qquad (6.115)$$

and this amount should be removed from the end of the open stub.

All of the above calculations were performed at the target frequency of 500 MHz. It is, however, always prudent to check stability at other frequencies for which the device exhibits significant gain. (Parasitic oscillations, especially at low frequencies, are a common problem in amplifier design.) At 100 MHz the transistor has typical S parameters

$$\left. \begin{matrix} S_{11} = 0.65\angle310°, & S_{21} = 8.2\angle123° \\ S_{12} = 0.028\angle85°, & S_{22} = 0.84\angle347° \end{matrix} \right\} Z_0 = 50\,\Omega$$

for which $K = 1.003$ and $|\Delta| = 0.59$. In addition, at 900 MHz, the transistor has typical S parameters

$$\left. \begin{matrix} S_{11} = 0.32\angle202°, & S_{21} = 1.6\angle53° \\ S_{12} = 0.11\angle58°, & S_{22} = 0.67\angle323° \end{matrix} \right\} Z_0 = 50\,\Omega$$

for which $K = 1.79$ and $|\Delta| = 0.18$. Between these frequencies, the amplifier remains unconditionally stable. At the lower frequency end, however, it will be noted that the gain is greatly increased and the stability conditions are only marginally satisfied. Fortunately, as the frequency drops further, the bias and supply lines will increasingly behave as short circuits to ground and so will serve to stabilise the amplifier.

6.8   The measurement of *S* parameters

We have shown how useful an S parameter matrix description can be, but it remains to be explained how the necessary S parameters can be measured in practice. To do this, we need the ability to separate out forward and reflected waves. A device that can achieve this decomposition is known as a directional coupler and an example is shown in Figure 6.27. The coupler is a four-port device with characteristic impedance Z_0 at all ports. The transformer winding ratio n is normally chosen large enough for the direct transmission from ports 1 to 2 to be little affected by the coupler. Let V^+ and V^- be the right and left travelling voltage waves in the line between ports 1 and 2. Assuming that the transformers are tightly coupled, and the self-inductance is large, they can be treated as ideal. The left-hand transformer will act as an ideal current source that forces current $(1/nZ_0)(V^+ - V^-)$ into the line joining ports 3 and 4. This in turn will cause a current of magnitude $(1/2nZ_0)(V^+ - V^-)$ to flow through the loads on ports 3 and 4. The right-hand transformer, however, will act as an ideal voltage source of magnitude $(1/n)(V^+ + V^-)$. This will cause a current $(1/2nZ_0)(V^+ + V^-)$ to flow through the load on port 3 and current $(-1/2nZ_0)(V^+ + V^-)$ through the load on port 4. As a consequence, the voltage at ports 3 and 4 will be V^+/n and $-V^-/n$, respectively. These voltages can be used to infer the right and left travelling voltage waves on the line joining ports 1 and 2 and hence the reflection coefficient Γ at the load Z_L.

If we are only interested in measuring $|\Gamma|$, we can use simple envelope detectors at ports 3 and 4 to obtain $|V^+|$ and $|V^-|$ and hence $|\Gamma| = |V^-|/|V^+|$. If we require the phase aspect of Γ, we will need to find the relative phases of V^+ and V^-. To measure both transmission and reflection parameters for a two-port network, we will need a set-up of the form shown in Figure 6.28. Note that the scattering matrix components S_{11} and S_{21} can be derived from

$$S_{11} = \frac{V_1^-}{V_1^+} \quad \text{and} \quad S_{21} = \frac{V_2^-}{V_1^+} \tag{6.116}$$

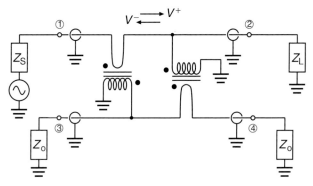

Figure 6.27   A simple directional coupler.

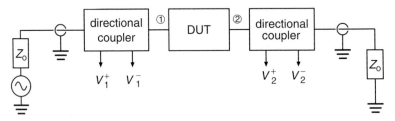

Figure 6.28   Set-up for the measurement of S parameters.

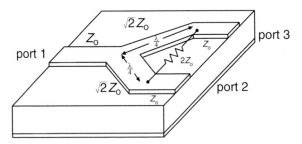

Figure 6.29   Microstrip form of Wilkinson power divider.

and the device under test (DUT) can be reversed in order to find S_{12} and S_{22}. Such a process is the basis of a *network analyser*, an extremely useful RF test instrument.

6.9   Some useful multiport networks

So far we have only considered networks with one or two ports, but the S matrix description can be usefully applied to networks with any number of ports. Figure 6.29 shows a microstrip Wilkinson power divider, an example of a three-port network. Power is fed into port 1 and then divided between the loads on ports 2 and 3 (evenly if the loads are equal). For equal matched loads, there is no flow of current across the divider resistance and hence no loss. If there is no current flow across the divider resistor, we can ignore this component and choose quarter-wave lines that transform each load impedance Z_{0} into $2Z_{0}$. These transformed loads combine at port 1 to present a load of impedance Z_{0} to the source at this port. An impedance transformation from Z_{0} to $2Z_{0}$ is achieved through a quarter-wave transmission line of impedance $\sqrt{2}Z_{0}$ and this is the reason for the quarter-wave lines in the divider. The S matrix of this three-port system is given by

$$S = \begin{pmatrix} 0 & -\dfrac{j}{\sqrt{2}} & -\dfrac{j}{\sqrt{2}} \\ -\dfrac{j}{\sqrt{2}} & 0 & 0 \\ -\dfrac{j}{\sqrt{2}} & 0 & 0 \end{pmatrix}. \tag{6.117}$$

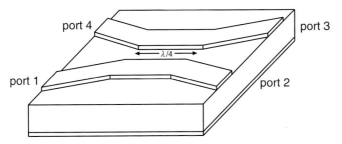

Figure 6.30   Microstrip form of directional coupler.

It will be noted that the matrix has zeros on the diagonal since the device is *matched* on all ports (this means that the impedance looking into any port is the characteristic impedance Z_o when all other ports are loaded with Z_o). Furthermore, the other two zeros in the matrix imply that the ports 2 and 3 are isolated.

We have already seen the utility of directional couplers and Figure 6.30 shows a microstrip realisation of such a device. For this realisation, the input into any one port will divide itself between the two adjacent ports. Clearly, there is direct transmission along one microstrip and some transference of power between the quarter-wave lines through their mutual capacitance and inductance. Crucially, however, there is negligible transference of power to the port that is diametrically opposite since the coupled fields in this direction will mutually cancel. The S matrix of the ideal coupler is given by

$$S = \begin{pmatrix} 0 & -j\sqrt{1 - c_{\lambda/4}^2} & 0 & c_{\lambda/4} \\ -j\sqrt{1 - c_{\lambda/4}^2} & 0 & c_{\lambda/4} & 0 \\ 0 & c_{\lambda/4} & 0 & -j\sqrt{1 - c_{\lambda/4}^2} \\ c_{\lambda/4} & 0 & -j\sqrt{1 - c_{\lambda/4}^2} & 0 \end{pmatrix} \tag{6.118}$$

which reflects the high degree of symmetry in this device. The coupling strength between the microstrips is measured in terms of the coefficient $c_{\lambda/4}$ and the coupling is expressed in terms of

$$C_{\lambda/4} = 10 \log \frac{P_1}{P_4} = 10 \log |S_{14}|^{-2} = -20 \log c_{\lambda/4} \text{ dB}, \tag{6.119}$$

where P_4 is the power that issues from port 4 when power P_1 enters port 1. No power should emerge from port 3 and deviations from this ideal are normally expressed in terms of the directivity

$$D_{\lambda/4} = 10 \log \frac{P_4}{P_3} \text{ dB}, \tag{6.120}$$

where P_3 is the power that issues from port 3.

The action of the coupler can be analysed in terms of *even* and *odd* modes. Consider parallel microstrips (see Figure 6.31a) that have self-capacitance C, self-inductance L,

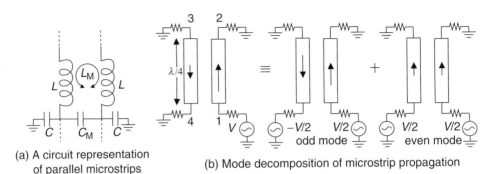

(a) A circuit representation
of parallel microstrips

(b) Mode decomposition of microstrip propagation

Figure 6.31   Analysis of microstrip coupler in terms of even and odd modes.

mutual capacitance C_M and a mutual inductance of L_M (all per unit length). We excite port 1 with a source of impedance Z_0 (the characteristic impedance) and load all other ports with Z_0. The propagation on the microstrips can be analysed as a combination of even (equal current) and odd (opposite current) transmission modes (see Figure 6.31b). For a particular mode, the propagation along one microstrip will need to take account of the current on the other microstrip. Because of this, the effective inductance per unit length in the odd mode will be $L - L_M$ and the effective capacitance $C + 2C_M$. In the case of the even mode, there will be effective inductance per unit length of $L + L_M$ and effective capacitance of C. As a consequence, there will be characteristic impedances $Z_e = \sqrt{(L + L_M)/C}$ and $Z_o = \sqrt{(L - L_M)/(C + 2C_M)}$ for the even and odd modes, respectively. For even and odd modes, the loads on ports 2 and 3 will transform down the quarter-wave transmission lines as Z_e^2/Z_0 and Z_o^2/Z_0 respectively. As a consequence, the even and odd mode reflection coefficients at ports 1 and 4 will be given by $\Gamma_e = (Z_e^2 - Z_0^2)/(Z_e^2 + Z_0^2)$ and $\Gamma_o = (Z_o^2 - Z_0^2)/(Z_o^2 + Z_0^2)$. If a voltage wave V^+ is incident upon port 1, this can be regarded as the sum of even and odd modes. A voltage wave of magnitude $(\Gamma_e - \Gamma_o)V^+/2$ will exit at port 4 (the sum of reflections for both even and odd modes) and a wave of magnitude $(\Gamma_e + \Gamma_o)V^+/2$ at port 1. It will be noted that $\Gamma_e + \Gamma_o = (Z_e^2 - Z_0^2)/(Z_e^2 + Z_0^2) + (Z_o^2 - Z_0^2)/(Z_o^2 + Z_0^2)$ and, if we choose a characteristic impedance of magnitude $Z_0 = \sqrt{Z_e Z_o}$, $\Gamma_e + \Gamma_o = 0$. Consequently, the device will be matched at port 1 and, by symmetry, at all other ports. In addition, $\Gamma_e - \Gamma_o = 2(Z_e - Z_o)/(Z_e + Z_o)$ and the power flowing out of port 4 will have a voltage of magnitude $V^+(Z_e - Z_o)/(Z_e + Z_o)$. As a consequence, the coupling factor will be given by $C_{\lambda/4} = 20\log|(Z_e + Z_o)/(Z_e - Z_o)|$ dB. Of great importance to the operation of the directional coupler is the output at port 3. From the considerations of Section 6.5, even mode voltage $(V^+/2j)[1/(1 - \Gamma_{3e}^2)]$ and odd mode voltage $-(V^+/2j)[1/(1 - \Gamma_{3o}^2)]$ will be incident upon the load at this port ($\Gamma_{3e} = (Z_0 - Z_e)/(Z_0 + Z_e)$ and $\Gamma_{3o} = (Z_0 - Z_o)/(Z_0 + Z_o)$ are the even and odd mode reflection coefficients at port 3, respectively). Since $Z_0 = \sqrt{Z_e Z_o}$, we will have $\Gamma_{3e} = -\Gamma_{3o}$ and the total output at port 3 will be zero (i.e., $S_{31} = 0$).

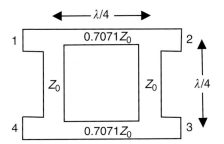

Figure 6.32   A branch line quadrature hybrid.

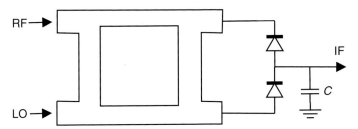

Figure 6.33   A microwave mixer that is based on a 90° hybrid and diode pair.

Figure 6.32 shows another very useful four-port device, the branch line quadrature hybrid. For this device, any power that enters ports 1 or 4 will be split evenly between ports 2 and 3. The output at ports 2 and 3 will, however, be out of phase by 90°. It should be noted that the microstrips that form the hybrid are all a quarter wavelength at the operating frequency. Furthermore, the horizontal microstrips should have an impedance $Z_0/\sqrt{2}$ and the vertical microstrips should have an impedance Z_0 (Z_0 is the required characteristic impedance of the device). The S matrix of the hybrid is given by

$$S = \begin{pmatrix} 0 & \frac{-j}{\sqrt{2}} & \frac{-1}{\sqrt{2}} & 0 \\ \frac{-j}{\sqrt{2}} & 0 & 0 & \frac{-1}{\sqrt{2}} \\ \frac{-1}{\sqrt{2}} & 0 & 0 & \frac{-j}{\sqrt{2}} \\ 0 & \frac{-1}{\sqrt{2}} & \frac{-j}{\sqrt{2}} & 0 \end{pmatrix} \tag{6.121}$$

from which it will be noted that ports 1 and 4 are isolated. Such hybrids can be extremely useful in the construction of microwave mixers, as shown in Figure 6.33. Unfortunately, this particular example has the disadvantage that both input signals (LO and RF) will be present at the IF output and hence necessitate some good output filtering. On the plus side, however, this form of mixer can greatly reduce the impact of local oscillator noise.

Figure 6.34a shows a circulator, another important three-port device. This device has the unique property that it can transmit power from ports 1 to 2, from ports 2 to 3 and from ports 3 to 1, but no other form of power transmission is possible. Under such

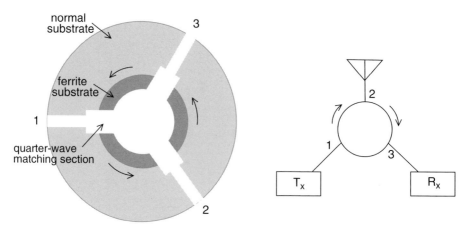

(a) Microstrip circulator   (b) Circulator used to connect a transceiver to its antenna

Figure 6.34   Microstrip circulator with application.

conditions, the S matrix will have the form

$$S = \begin{pmatrix} 0 & 0 & 1 \\ 1 & 0 & 0 \\ 0 & 1 & 0 \end{pmatrix}. \tag{6.122}$$

It will be noted that the ports of the device join together on a central conducting disc under which the substrate has been replaced by a ferrite material. This ferrite material is magnetised by a static magnetic field that is perpendicular to the plane of the device (this field is usually produced by a permanent magnet that is placed under the ground plane). By correctly adjusting the magnetic field, it is possible to set up electromagnetic modes within the ferrite that couple into adjacent ports in one direction and not the other. (This non-reciprocal behaviour is typical of devices that are based on ferrite materials.) Circulators can be extremely useful devices in tranceivers (transmitter/receiver systems) that need to use a single antenna for both receive and transmit functions, as illustrated in Figure 6.34b.

6.10   Reflection coefficient approach to microwave oscillators

Reflection coefficients can provide an alternative means of analysing oscillators and the approach is particularly useful at microwave frequencies (the gigahertz range). Consider the amplifier configuration shown in Figure 6.35. We need the amplifier to be unstable for oscillations to occur and, in terms of the reflection coefficients, this will require $|\Gamma_{in}| > 1$ and/or $|\Gamma_{out}| > 1$. Z_L could be chosen such that $|\Gamma_{in}| > 1$ (note that Γ_{in} depends on Γ_L) and Z_S chosen such that $Z_S + Z_{in} = 0$ at the frequency for which oscillation is required. (The initial input voltage that starts the oscillation is generated

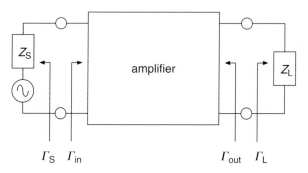

Figure 6.35   Amplifier configuration.

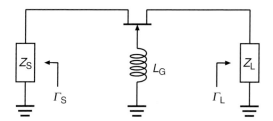

Figure 6.36   An FET negative resistance oscillator.

by the noise in the circuit.) Since $Z_S = -Z_{in}$ at the frequency of oscillation

$$\Gamma_S = \frac{Z_S - Z_0}{Z_S + Z_0} = \frac{-Z_{in} - Z_0}{-Z_{in} + Z_0} = \frac{1}{\Gamma_{in}} \tag{6.123}$$

and so

$$\Gamma_S \Gamma_{in} = 1 \tag{6.124}$$

(this is the equivalent of the Barkhausen criterion). Note that $\Gamma_S \Gamma_{in} = 1$ will also imply that $\Gamma_L \Gamma_{out} = 1$.

Most transistors are designed to be stable and so a major problem in oscillator design is the manufacture of a suitable instability. A common-gate FET (or common-base BJT) amplifier can be made unstable by adding some series inductance to the gate (or base in the case of a BJT). An example of a common-gate FET oscillator is given in Figure 6.36 (bias not shown). Assuming an FET model that consists of a current source and non-zero C_{GS} (we take $C_{DS} = C_{GD} = 0$ and $r_d = r_\pi = \infty$), the common-gate amplifier input admittance (see Figure 6.37) will be of the form

$$Y_{in} = \frac{j\omega C_{GS} + g_m}{1 - \omega^2 L_G C_{GS}} \tag{6.125}$$

which is independent of the load. Above the resonant frequency of L_G and C_{GS} is series, this admittance will have a negative real part and so the amplifier will be unstable. The input reflection coefficient will be given by $\Gamma_{in} = (Y_0 - Y_{in})/(Y_0 + Y_{in})$ where $Y_0 = Z_0^{-1}$ and from which it can be seen that $|\Gamma_{in}| > 1$ above the resonant frequency. At very

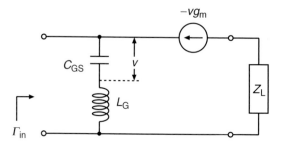

Figure 6.37   Model of FET negative resistance generator.

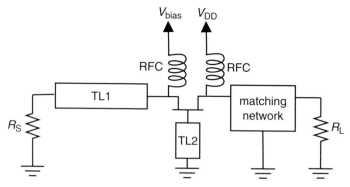

Figure 6.38   Microstrip implementation of a negative resistance oscillator.

high frequencies, the inductance of the transistor leads can sometimes be sufficient to cause such an instability. This fact serves to emphasise the need for the RF designer to take full account of the non-ideal nature of components, and their connections, when attempting to design an amplifier and not an oscillator.

To design an oscillator we need to choose a combination of inductor L_G and load Z_L such that $\Gamma_{in} > 1$ (it is usual to choose Z_L real in a first-cut design). Impedance Z_S ($= R_S + jX_S$) is now chosen such that $X_S = -X_{in}$ in order to resonate the circuit. It should be noted, however, that choosing $R_S = -R_{in}$ is often not appropriate since, as the amplitude of oscillation rises, R_{in} tends to become less negative. A good rule of thumb is to choose $R_S = -R_{in}/3$, where R_{in} is the small signal input resistance of the amplifier. The oscillation level will rise until equilibrium is achieved through gain compression. Figure 6.38 shows a microstrip implementation of a negative resistance oscillator. Transmission line TL2 in the gate provides the necessary inductance to make the FET unstable and the transmission line TL1, together with resistor R_S, presents a suitable impedance to the FET source. The oscillator load is connected to the drain through a matching network (a microstrip L-network, for example) and the bias and drain supplies are connected through RF chokes. As in the earlier amplifier design (Section 6.7), these chokes can be replaced by RF-shorted quarter-wave microstrips.

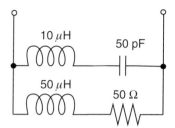

Figure 6.39   Circuit for Smith chart example in Question 5.

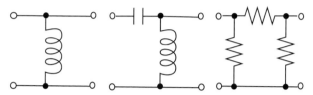

Figure 6.40   Circuits for two-port S matrix calculations in Question 6.

EXERCISES

(1) Design a 50 Ω coaxial cable with less than 0.01 dB loss per metre at 1 GHz. The dielectric should be a vacuum ($G = 0$) and the conductor should be made of copper ($R = (R_s/2\pi)(1/a + 1/b)$, where $R_s = \sqrt{\omega\mu/2\sigma}$ is the surface resistance and $\sigma = 5.8 \times 10^7$ S/m is the conductively of copper).

(2) For FR4 PCB material ($\epsilon_r = 4.4$ and $\delta = 0.01$) with substrate thickness 3 mm, find the width of microstrip (assumed to be made of copper) that yields a 50 Ω characteristic impedance. Find the loss per metre at a frequency of 2 GHz.

(3) Design a quarter-wave microstrip transformer that matches a 200 Ω load into a 50 Ω microstrip transmission line (assume a substrate with $\epsilon_r = 4.4$ and thickness 1.5 mm).

(4) Write a MATLAB program (or with other suitable software) that calculates how a general impedance Z_L is transformed down a transmission line of impedance Z_o and propagation constant β. Modify this program to calculate the length of transmission line and stub that will transform the load impedance Z_L into the impedance Z_T of a connecting transmission line. Use this program to find the matching network that will transform the load $Z_L = 20 + j10 \Omega$ into 50 Ω (use a 50 Ω transmission line with on capacitance of 110 pF/m). Calculate the discrete component that could be used to replace the stub.

(5) Calculate the reflection coefficients of the circuit shown in Figure 6.39 for a range of frequencies that includes the circuit resonances. Display this behaviour on a Smith chart and identify the nature of the circuit resonances.

(6) Calculate the S and Z matrices for the circuits shown in Figure 6.40. Assume a general value L for inductors, a general value C for capacitors and value Z_o (the characteristic impedance) for the resistors.

(7) The two-port networks A and B have scattering matrices S^A and S^B, respectively. They are connected in series (port 2 of network A to port 1 of network B) to form a new two-port network with scattering matrix S. Show that

$$S = \begin{pmatrix} s_{11}^A + \dfrac{s_{12}^A s_{21}^A s_{11}^B}{1 - s_{11}^B s_{22}^A} & \dfrac{s_{12}^A s_{12}^B}{1 - s_{11}^B s_{22}^A} \\[3mm] \dfrac{s_{21}^B s_{21}^A}{1 - s_{11}^B s_{22}^A} & s_{22}^B + \dfrac{s_{21}^B s_{12}^B s_{22}^A}{1 - s_{11}^B s_{22}^A} \end{pmatrix}.$$

(8) Design a maximum gain amplifier for 300 MHz with 50 Ω source and load impedances. Use a Philips BFR53 semiconductor (assume $S_{11} = 0.3\angle155°$, $S_{21} = 9\angle85°$, $S_{12} = 0.05\angle60°$ and $S_{22} = 0.45\angle335°$ at 300 MHz) and LC-matching networks. Redesign the input circuit for minimum noise ($F_{min} = 3.3$ dB and optimum admittance $Y_{opt} = 18 - j4$ mS).

SOURCES

Akhtarzad, S., Rowbotham, T. R. and Johns, P. B. 1975. The design of coupled microstrip lines. In *IEEE Transactions on Microwave Theory and Techniques*, Vol. MTT-23.

Bowick, C. 1982. *RF Circuit Design*. Boston, MA: Newnes.

Bryant, G. H. 1993. *Principles of Microwave Measurements* (revised ed.). IEE Electrical Measurement series, Volume 5. London: Peter Peregrinus.

Collin, R. E. 1992. *Foundations for Microwave Engineering* (2nd edn). New York: McGraw-Hill.

Fooks, E. H. and Zakarevicins, R. A. 1990. *Microwave Engineering Using Microstrip Circuits*. New York: Prentice-Hall.

Hall, G. J. (ed.). 1988. *The ARRL Antenna Book*. Newark, CT: American Radio Relay League.

Ludwig, R. and Bretchko, P. 1990. *RF Circuit Design: Theory and Application*. Upper Saddle River, NJ: Prentice-Hall.

Hayward, W. 1994. *Introduction to Radio Frequency Design*. Newark, CT: American Radio Relay League.

Popovic, Z. and Popovic, B. D. 2000. *Introductory Electromagnetics*. Upper Saddle River, NJ: Prentice-Hall.

Pozar, D. M. 1998. *Microwave Engineering* (2nd edn). New York: John Wiley.

Pozar, D. M. 2001. *Microwave and RF Design of Wireless Systems*. New York: John Wiley.

Sabin, W. E. and Schoenike, E. O. (eds.). 1998. *HF Radio Systems and Circuits*. Atlanta, GA: Noble Publishing Corporation.

Smith, A. A. 1998. *Radio Frequency Principles and Applications*. Piscataway, NJ: IEEE Press.

Wheeler, H. A. 1977. Transmission line properties of a strip on a dielectric sheet on a plane. In *IEEE Transactions on Microwave Theory and Techniques*, Vol. 25, pp. 631–647.

7 Power amplifiers

RF signals will often need to be transmitted with considerable power if they are to survive propagation with adequate signal level. As a consequence, we will need to consider amplifiers that can operate at large signal levels. Up to this point, we have concentrated on small signal amplifiers for which efficiency and linearity have not been a major problem. These aspects, however, require careful consideration in the case of RF amplifiers operating at large signal levels. Small signal amplifiers are typically of the class A variety and highly linear. Whilst class A amplifiers are sometimes used at high power levels, they do not represent an efficient use of the d.c. energy that is supplied to the amplifier. Class B, AB, C and E amplifiers are far more efficient, but have the disadvantage that they are highly non-linear and hence create considerable harmonics. These harmonics can be troublesome and require specialised techniques, or filtering, for them to be brought down to an acceptable level. The following chapter considers power amplifiers in the class range from A to E. It concentrates on BJT amplifiers, but the same principles can be applied to FET amplifiers.

7.1 Class A

Class A amplifiers attempt to operate over that part of the transistor characteristic for which there is linear translation of the input signal to the output. For a BJT, a typical configuration is shown in Figure 7.1. The transistor is biased such that it conducts throughout the cycle (conduction angle of 360°) and gives linear amplification throughout its operation (this is illustrated in Figure 7.2). We normally require a design that gives maximum output power, not necessarily maximum gain, since large amounts of gain are better achieved at low power levels. Consequently, we choose a load R_L that gives maximum voltage swing at the output. Since the collector voltage v_{CE} is related to the output voltage v_o through $v_{CE} \approx V_{CC} + v_o$, it is obvious that the maximum possible excursion of the output RF voltage v_o will be V_{CC} if the transistor is to remain in conduction throughout. Furthermore, if I_{DC} is the d.c. part of the collector current and i_{RF} the RF part, the maximum possible excursion of i_{RF} will be I_{DC}. The output current i_o will satisfy $i_o = -i_{RF}$ since no RF current can flow through the RF choke

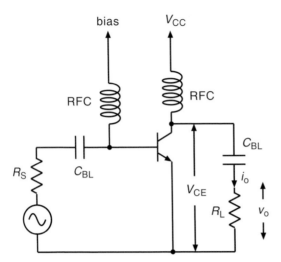

Figure 7.1   Basic power amplifier circuit.

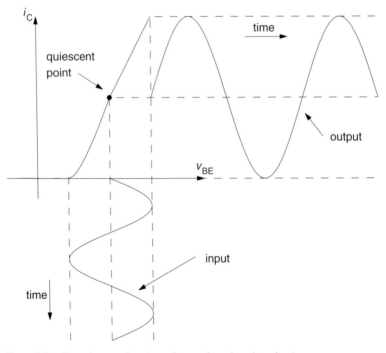

Figure 7.2   Class A operation (waveforms plotted against time).

and, as a consequence, $v_o = -R_L i_{RF}$. It is clear that we will need to set the bias such that $I_{DC} \approx V_{CC}/R_L$ in order to achieve maximum swing. For a sinusoidal drive, the average output power P_o will be limited according to

$$P_o \leq \frac{V_{CC}^2}{2R_L} \qquad (7.1)$$

and the d.c. input power will be given by $P_{DC} = V_{CC} I_{DC} = V_{CC}^2/R_L$. Consequently,

$$\text{efficiency} = \frac{P_o}{P_{DC}} \leq 0.5 \text{ (or 50\%)}. \tag{7.2}$$

The power dissipated in the transistor $P_T (= P_{DC} - P_o)$ will take its maximum value V_{CC}^2/R_L when there is no drive. For a strictly linear amplifier, it is not advisable to have a maximum voltage excursion of V_{CC} since this will lead to the transistor operating outside the active region (the region where the input and output currents are linearly related). In practice, we will need to limit voltage excursions to $V_{CC} - V_{sat}$, where V_{sat} is the saturation voltage (the collector voltage at which the active region starts).

In designing a power amplifier, we will normally be restricted by the supply voltage V_{CC} and output power P_o. If the amplifier is required to deliver power P_o, then a load of $R_L = (V_{CC} - V_{sat})^2/2P_o$ will be required to achieve the maximum acceptable swing. If the given load is different from this, we will need to add a matching network to transform it into R_L. L- or π-matching networks are often used since they can also provide a trap for unwanted harmonics. The transistor will need to be biased for a quiescent current of $I_{DC} = V_{CC}/R_L$ and the amplifier signal source will normally need to be matched to the input impedance of the transistor. In choosing a transistor, we need to make sure that it can handle a maximum current of $2V_{CC}/R_L$, a maximum voltage of $2V_{CC}$ and can dissipate a power of at least V_{CC}^2/R_L.

7.2   Class B

In class B amplifiers, the transistor is biased such that it only conducts for half a cycle (conduction angle of 180°). The operation is far more efficient than class A, but suffers from a large amount of distortion on the output signal. Fortunately, this class of amplifier lends itself to techniques that can remove much of the distortion whilst retaining the high efficiency. The relationship between the input and output signals is illustrated in Figure 7.3. We will approximate the collector current by

$$i_C(t) = \begin{cases} I_P \sin \omega t & 0 \leq \omega t \leq \pi \\ 0 & \pi \leq \omega t \leq 2\pi, \end{cases} \tag{7.3}$$

(where I_P represents the peak current). The d.c. current is given by

$$I_{DC} = \frac{1}{T} \int_0^T i_C(t)\, dt = \frac{I_P}{\pi}, \tag{7.4}$$

where $T = 2\pi/\omega$ is the period. Unfortunately, the collector current will contain components at frequencies other than the fundamental frequency and these will

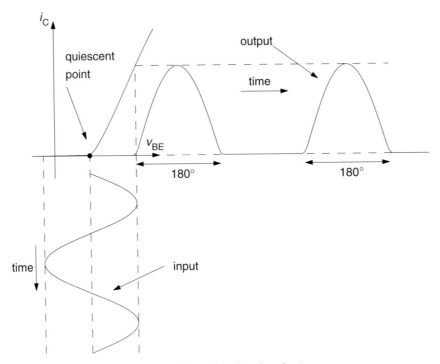

Figure 7.3   Class B operation (waveforms plotted against time).

normally be filtered out. Consequently, we will be concerned with the fundamental component of current $(I_P/2) \sin \omega t$ and for which there is an average output power

$$P_o = \frac{I_P^2}{8} R_L. \tag{7.5}$$

Furthermore, the d.c. power supplied is given by

$$P_{DC} = V_{CC} I_{DC} = \frac{V_{CC} I_P}{\pi}. \tag{7.6}$$

From these values,

$$\text{efficiency} = \frac{P_o}{P_{DC}} = \frac{\pi I_P R_L}{8 V_{CC}}. \tag{7.7}$$

As with the class A amplifier, the maximum possible excursion of the RF output voltage will be V_{CC} and hence $P_o \leq V_{CC}^2/2R_L$ which, from Equation 7.5, will imply $I_P \leq 2V_{CC}/R_L$ and hence a maximum efficiency of $\frac{\pi}{4}$ (or 78.5 %). The dissipation in the transistor is given by

$$P_T = P_{DC} - P_o = \frac{I_P V_{CC}}{\pi} - \frac{I_P^2 R_L}{8}, \tag{7.8}$$

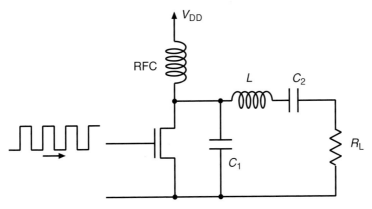

Figure 7.7 Class E amplifier.

before the transistor is next switched on. It can be shown (see Sokal and Sokal) that the appropriate values for C_1, C_2 and L can be calculated according to

$$C_1 \approx \frac{1}{5.447\omega R_L}, \quad L = \frac{QR}{\omega} \quad \text{and} \quad C_2 \approx C_1\frac{5.447}{Q}\left(1 + \frac{1.42}{Q - 2.08}\right),$$

where $R_L = 0.577(V_{DD}/P_o)$, P_o is the desired output power and Q is chosen to yield the desired bandwidth. If R_L is not equal to the desired load, an additional output matching network will be required.

7.5   A design example

Example Design a 150 MHz class C amplifier that uses an 8 °V power supply and provides 2.0 W into a 50 Ω load from a 50 Ω source.

We choose an SGS-Thomson SD1135-03 bipolar device since this can handle up to 2.5 W with 11 dB of gain. Typical input and output impedances of this device at 150 MHz are

$$Z_{in} = 2.2 - j0.4 \ \Omega \quad \text{and} \quad Z_{out} = 7.9 + j8.4 \ \Omega.$$

Since the supply voltage and output power have been specified, we need to choose R_L according to

$$R_L = \frac{V_{CC}^2}{2P_o} = \frac{8^2}{2 \times 2} = 16 \ \Omega.$$

It is clear that we will require an output network that transforms the 50 Ω of the specified load into 16 Ω and also cancels the 8.4 Ω of inductive impedance at the transistor output. At the input, we will need to transform the input impedance Z_{in} of the transistor into the 50 Ω of the source.

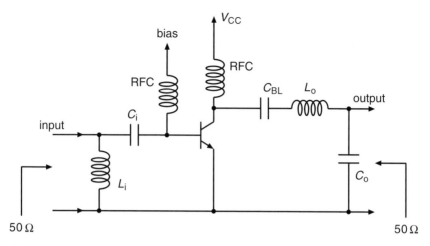

Figure 7.8   Power amplifier example.

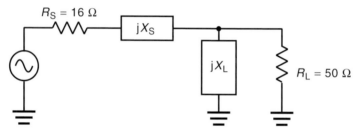

Figure 7.9   Design of an output matching network.

Before proceeding with the design, we need to check that the dissipation and maximum current are within the specification of the transistor. The worst case (class B)
will cause a dissipation of 0.8 W which is well within the capacity of the transistor.
The maximum current that an SD1135-03 can handle will be of the order of 1.7 A and,
noting Equation 7.17, the values of θ will need to satisfy

$$\frac{2\pi \times 8}{16} \frac{1 - \cos \theta}{2\theta - \sin 2\theta} < 1.7. \tag{7.18}$$

This will require the conduction angle 2θ to be greater than 84.8° (the angle is set by
the bias). From Equation 7.15, this will imply an efficiency of less than 94 per cent.

Figure 7.8 shows a possible amplifier design. The output L-network provides the
requisite impedance transformation of the load and also acts to filter out frequencies
above the 150 MHz operating frequency (this removes any harmonics that are created
by the non-linearities in the transistor). As shown in Figure 7.9, the required output
network is one which can match a 16 Ω source into a 50 Ω load. The network reactances

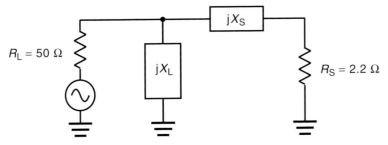

Figure 7.10   Input matching network.

can be calculated from the design formulas

$$X_L = -\frac{R_L}{Q} \quad \text{and} \quad X_S = -\frac{X_L Q^2}{Q^2 + 1}, \tag{7.19}$$

where $Q = \sqrt{(R_L/R_S) - 1}$. Since $Q = 1.46$, we will have $X_L = -34.25 \ \Omega$ and $X_S = 23.33 \ \Omega$. Note that X_S must also include a reactance of $8.4 \ \Omega$, contributed by the transistor output, so the reactance of L_o in Figure 7.8 will need to be $14.93 \ \Omega$. Since $\omega = 2\pi \times 1.5 \times 10^8$ rad/s and $\omega L_o = 14.93$, L_o will have a value of 15.8 nH. Furthermore, since $X_L = -1/\omega C_o$, C_o in Figure 7.8 will have the value 31 pF.

At the amplifier input, there is an L-network that transforms the $50 \ \Omega$ source impedance into $2.2 \ \Omega$. This network is illustrated in the Figure 7.10 (it will be noted that we have reversed the prototype network of Chapter 2 since the larger impedance is that of the source). From the network design formulae, we find that $Q = 4.66$ which implies $X_L = 10.7 \ \Omega$ and $X_S = -10.2 \ \Omega$. Note that X_S must include a reactance of $-0.4 \ \Omega$ that is contributed by the transistor input. Consequently, the reactance of C_i in Figure 7.8 will need to be $-9.8 \ \Omega$. The final values for the input network components will be $C_i = 108$ pF and $L_i = 10.8$ nH.

7.6   Transmission line transformers

In all our considerations thus far, we have concentrated on matching circuits that perform their required function at, or close to, a single frequency. There are, however, circumstances where we need an amplifier to operate over a range of frequencies and the normal tuned transformers are not appropriate. Transmission line transformers (see Ruthroff or Smith) offer an alternative for such situations. The operation of these transformers can be illustrated through the circuit of Figure 7.11. For the transmission line (see Equations 6.28 and 6.29 in Chapter 6), we have

$$v_1 = v_2 \cos \beta l + j i_2 Z_o \sin \beta l \tag{7.20}$$

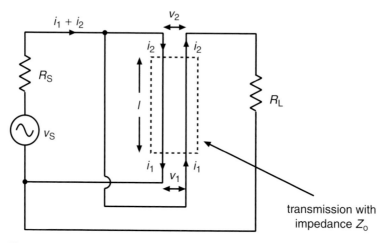

Figure 7.11   Transmission line transformer model.

and

$$i_1 = i_2 \cos \beta l + \frac{jv_2}{Z_o} \sin \beta l. \tag{7.21}$$

For voltage balance around the circuit,

$$v_S = (i_1 + i_2)R_S + v_1 \tag{7.22}$$

and

$$v_1 + v_2 = i_2 R_L. \tag{7.23}$$

Eliminating v_1 and v_2 between Equations 7.20 to 7.23, we obtain

$$v_S - (i_1 + i_2)R_S = [(i_1 + i_2)R_S - v_S + i_2 R_L] \cos \beta l + j i_2 Z_o \sin \beta l \tag{7.24}$$

and

$$i_1 = i_2 \cos \beta l + \frac{j}{Z_o}[(i_1 + i_2)R_S - v_S + i_2 R_L] \sin \beta l. \tag{7.25}$$

Then, eliminating i_1 between Equations 7.24 to 7.25 (some tortuous algebra),

$$i_2 = \frac{v_S(1 + \cos \beta l)}{2R_S(1 + \cos \beta l) + R_L \cos \beta l + j(R_S R_L + Z_o^2)\left(\frac{\sin \beta l}{Z_o}\right)} \tag{7.26}$$

from which the output power will be given by

$$P_o = \frac{1}{2}|i_2|^2 R_L$$

$$= \frac{1}{2}\frac{|v_S|^2(1 + \cos \beta l)^2 R_L}{(2R_S(1 + \cos \beta l) + R_L \cos \beta l)^2 + \left(R_S R_L + Z_o^2\right)^2 \left(\frac{\sin \beta l}{Z_o}\right)^2}. \tag{7.27}$$

Figure 7.12 Twisted pair transmission line.

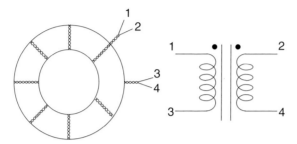

Figure 7.13 Practical transformer.

For a fixed source impedance R_S, the output power will be maximised with respect to Z_o when $Z_o = \sqrt{R_S R_L}$. Then, for this value of Z_o,

$$P_o = \frac{1}{2} \frac{|v_S|^2 (1 + \cos \beta l)^2 R_L}{[2R_S(1 + \cos \beta l) + R_L \cos \beta l]^2 + 4 R_S R_L \sin^2 \beta l} \tag{7.28}$$

which, for small βl, yields

$$P_o = \frac{1}{2} \frac{4|v_S|^2 R_L}{(4 R_S + R_L)^2}. \tag{7.29}$$

This power will then be the maximum available when $R_L \approx 4 R_S$ (i.e., the transformer matches source R_S to load $4 R_S$).

For a transmission line transformer, power transfer falls away as βl increases until it reaches a value of 0 when $\beta l = \pi$. Consequently, we normally design the transformer so that $l < \lambda/10$ at the highest operating frequency. At lower frequencies the transmission line model breaks down and the configuration acts more like an inductive transformer. Consequently, to improve performance at lower frequencies, the inductance is usually increased by the addition of a high permeability core. A typical transformer consists of a twisted pair transmission line (Figure 7.12) that is wound around a toroidal core (Figure 7.13). (The twisted pair will have $Z_o \approx 50\,\Omega$ for 2.5 mm diameter wire and one twist per cm down to $Z_o \approx 30\,\Omega$ for 5 twists per cm (see ARRL Handbook).) Note that Z_o (the characteristic impedance of transmission line) should be chosen such that $Z_o = \sqrt{R_S R_L}$ in order to achieve maximum power transfer. Figures 7.14 to 7.16 show several configurations that use transmission line transformers. Figures 7.14 and 7.15 show transformers that provide 4:1 and 9:1 impedance transformations, respectively, between source and load, whilst Figure 7.16 shows a power combiner (or alternatively a splitter) that could be used to implement a push-pull amplifier.

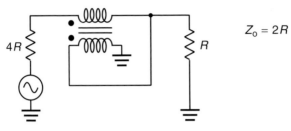

Figure 7.14   4:1 transformer.

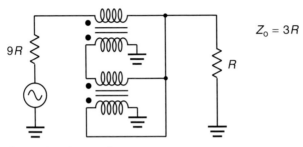

Figure 7.15   9:1 transformer.

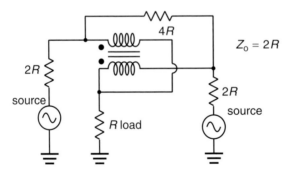

Figure 7.16   Power combiner.

A transmission line transformer can also be configured as a BALUN, a 1:1 trans-former that merely performs the function of connecting balanced and unbalanced cir-cuits (Figure 7.17). The distinction between these two types of circuit is important in that their direct connection can lead to strife. Simply, an unbalanced circuit is one for which the return path is through the ground or ground conductor. Examples are the microstrip line for which the return is through the ground plane and the coaxial cable for which the return is through the cable outer conductor. Unbalanced circuits are usually distin-guished by a distinct lack of geometric symmetry between the forward and return paths. A balanced circuit is one in which there is no ground return and is distinguished by its symmetry (the parallel wire transmission line, for example). The symmetric halves of

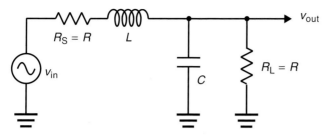

Figure 8.4 Two-pole filter example.

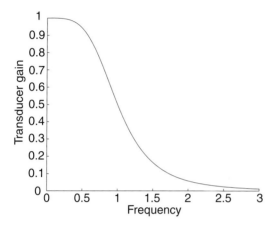

Figure 8.5 Transducer gain for two-pole filter (frequency scaled on the 3 dB frequency).

and $R \parallel (1/sC)$, we obtain the transfer function

$$H(s) = \frac{\left(sC + \frac{1}{R}\right)^{-1}}{R + sL + \left(sC + \frac{1}{R}\right)^{-1}}$$

$$= \frac{1}{s^2 LC + s\left(RC + \frac{L}{R}\right) + 2} \tag{8.5}$$

which has two poles

$$p_{1,2} = \frac{-\left(RC + \frac{L}{R}\right) \pm \sqrt{\left(RC + \frac{L}{R}\right)^2 - 8LC}}{2LC} \tag{8.6}$$

and no zeros. Figure 8.5 shows the frequency dependence of this filter in terms of transducer gain, the power delivered to the load ($v_{out}^2/2R_L$) when divided by the maximum power available from the source ($v_{in}^2/8R_S$). (Note that the frequency is scaled upon that for which the gain is 3 dB below the maximum.)

Typical transducer gain characteristics of a realistic low-pass filter are shown in Figure 8.6 and this illustrates some important factors in filter design. It is clear that an acceptable level of quantities such as passband width (dictated by the 3 dB bandwidth $\omega_{3\,dB}$), passband *ripple*, *insertion loss* and out of band *rejection* will need to

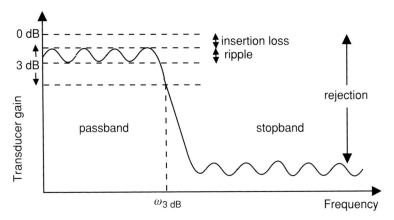

Figure 8.6 Typical low-pass filter characteristics.

be determined before the filter can be designed. The sharpness of the filter, normally defined in terms of the shape factor $SF = \omega_{60\,dB}/\omega_{3\,dB}$, will also need to be determined. Filter design consists of choosing poles and zeros to achieve the desired transfer function to a specified degree of accuracy. The transfer function will then be realised using lumped components or, as we shall see later, sections of transmission line.

For ease of design, simple procedures have been developed for some important subclasses of filter. Two important subclasses are the Chebyshev and Butterworth filters. For the Chebyshev low-pass filter, the transfer function has amplitude

$$|H_C(j\omega)| = \begin{cases} \dfrac{1}{\sqrt{1 + \epsilon^2 \cos^2\left[K \cos^{-1}\left(\frac{\omega}{\omega_c}\right)\right]}} & \left(\dfrac{\omega}{\omega_c}\right) < 1 \\[4mm] \dfrac{1}{\sqrt{1 + \epsilon^2 \cosh^2\left[K \cosh^{-1}\left(\frac{\omega}{\omega_c}\right)\right]}} & \text{otherwise} \end{cases} \tag{8.7}$$

which has the property that the ripples are of equal amplitude within the passband (parameter ϵ controls the amplitude of these ripples). In the case of the Butterworth filter, the transfer function has amplitude

$$|H_B(j\omega)| = \dfrac{1}{\sqrt{1 + \left(\frac{\omega}{\omega_c}\right)^{2K}}} \tag{8.8}$$

and is maximally flat in the passband. For both filters, K is the number of circuit elements (the number of inductors and capacitors) and ω_c is the 3 dB bandwidth (the frequency at which the transducer gain has fallen to half of its maximum). If passband ripples are acceptable, the Chebyshev filter will usually achieve a much better performance than the Butterworth filter with the same number of circuit components (Figure 8.7 compares these two filters for the same number of filter elements). A downside of the

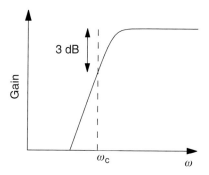

Figure 8.14   High-pass filter characteristics.

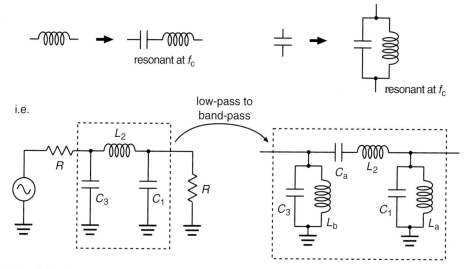

Figure 8.15   Band-pass filter derived from a low-pass filter.

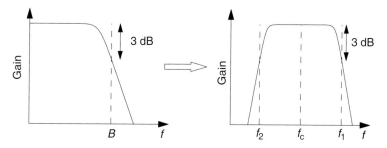

Figure 8.16   Transformation of filter characteristics.

f_c. An inductor L is resonated by a series capacitor $C = 1/L\omega_c^2$ and a capacitor C by a parallel inductor $L = 1/C\omega_c^2$ (note that $\omega_c = 2\pi f_c$). This is illustrated in Figure 8.15 with a typical frequency response shown in Figure 8.16. At the resonant frequency, the filter will act as a straight through connection between source and load. Filter response will fall away either side of frequency f_c with 3 dB points a frequency B apart.

Table 8.5 *Performance of a band-pass filter*

Frequency (MHz)	Transducer gain	Frequency (MHz)	Transducer gain
12.0	0.0005	15.0	1.0000
13.0	0.0029	15.1	0.9886
14.0	0.0488	15.2	0.9395
14.5	0.4293	15.3	0.8245
14.7	0.8151	15.5	0.4589
14.8	0.9374	16.0	0.0624
14.9	0.9885	17.0	0.0049

Example   Based on a three element low-pass filter, design a 1 MHz bandwidth band-pass filter with centre frequency 15 MHz and 50 Ω terminations.

We first design a low-pass filter with bandwidth of 1 MHz. Consider the low-pass Butterworth filter design with $K = 3$, then $L_2 = 2\,\mathrm{H}$ and $C_{1,3} = 1\,\mathrm{F}$ for 1 Ω terminations and 1 rad/s bandwidth. Rescaling for a 50 Ω load and 1 MHz bandwidth:

$$L_2 = \frac{50}{2\pi \times 10^6} \times 2\,\mathrm{H} = 15.915\,\mu\mathrm{H}$$

$$C_{1,3} = \frac{1}{2\pi \times 10^6 \times 50}\,\mathrm{F} = 3.1831\,\mathrm{nF}.$$

In order to resonate the low-pass filter elements at 15 MHz, we require

$$C_\mathrm{a} = \frac{1}{(2\pi f_\mathrm{c})^2 L_2} = 7.0736\,\mathrm{pF}$$

$$L_{\mathrm{a,b}} = \frac{1}{(2\pi f_\mathrm{c})^2 C_{1,3}} = 0.035\,37\,\mu\mathrm{H}$$

and the filter design is complete. Table 8.5 shows the performance of the filter in terms of its transducer gain.

8.4   Conversion of filters to microstrip form

Consider the low-pass filter shown in Figure 8.17. It is possible to approximate the capacitors and inductors by means of open and closed transmission line stubs, as shown in Figure 8.18. In this figure, the quantity Z_0 is the characteristic impedance of the transmission line. It is normal to design using standard $\lambda/8$ length stubs (λ is the transmission line wavelength at the 3 dB frequency ω_c of the desired filter). Consequently, an inductance L will require a shorted sub with $Z_0 = \omega_\mathrm{c} L$ and a capacitance C will require an open stub with $Z_0 = 1/\omega_\mathrm{c} C$. Away from ω_c, the behaviour of the $\lambda/8$ transmission lines will not exactly reproduce that of the corresponding discrete component, but it will be sufficiently close to produce a filter with the desired properties. Converted

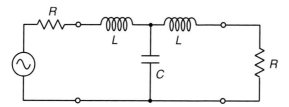

Figure 8.17   Circuit with three-element filter.

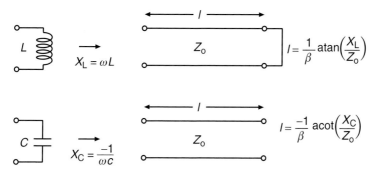

$$l = \frac{1}{\beta} \operatorname{atan}\left(\frac{X_L}{Z_0}\right)$$

$$l = \frac{-1}{\beta} \operatorname{acot}\left(\frac{X_C}{Z_0}\right)$$

Figure 8.18   Correspondence between lumped components and transmission lines.

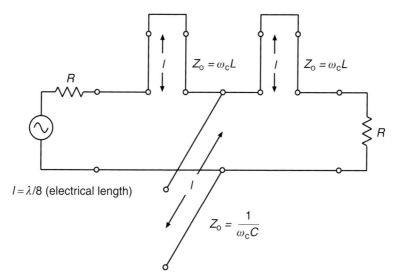

Figure 8.19   Basic three-element filter.

to transmission lines, the three-element filter of Figure 8.17 will take the form shown in Figure 8.19. Unfortunately, whilst the open-circuit stubs are easy to implement in microstrip technology, the shorted stubs are difficult. There is, however, a way around the problem. By means of Kuroda's identities (see Ludwig and Bretchko), it is possible transform the circuits into a collection of open transmission lines. The identities are

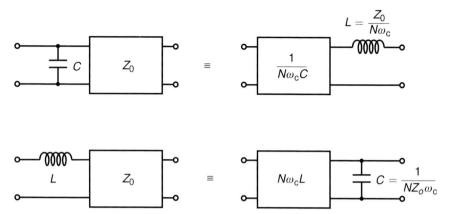

Figure 8.20 Some of Kuroda's identities.

illustrated in Figure 8.20 where $N = 1 + (1/\omega_c C Z_0)$ for the shunt capacitance and $N = 1 + (\omega_c L/Z_0)$ for the series inductance (the boxes are $\lambda/8$ transmission lines possessing the characteristic impedance shown in the box). The Kuroda identities can be verified at the 3 dB frequency ω_c by the application of the formulas for transforming impedance down a $\lambda/8$ transmission line, that is

$$Z_{in} = Z_0 \frac{Z_L + jZ_0}{Z_0 + jZ_L}, \tag{8.16}$$

where Z_{in} is the impedance looking into the transmission line, Z_0 is the characteristic impedance of the line and Z_L is the load on the line. Let the first circuit of the top identity in Figure 8.20 have load Z_L, then the input impedance is given by

$$Z_{in} = \left(\frac{1}{j\omega_c C}\right) \parallel \left(Z_0 \frac{Z_L + jZ_0}{Z_0 + jZ_L}\right)$$

$$= \frac{Z_0(Z_L + jZ_0)}{Z_0 + jZ_L + j\omega_c C Z_0(Z_L + jZ_0)}$$

$$= \frac{1}{N\omega_c C} \frac{Z_L + j\omega_c L + \frac{j}{N\omega_c C}}{\frac{1}{N\omega_c C} + j(Z_L + j\omega_c L)},$$

where $L = Z_0/N\omega_c$. The last expression for Z_{in} will be recognised as the series combination of inductance L and load Z_L when transformed along a $\lambda/8$ length transmission line of characteristic impedance $1/N\omega_c C$. As a consequence, we have verified the first Kuroda identity for the 3 dB frequency ω_c (the second identity can be verified in a similar fashion).

The filter section of Figure 8.17 is equivalent to the circuit shown in Figure 8.21 where R is the termination impedance of the filter. (Note that, since the additional transmission lines are matched to source and load, they have no effect upon the performance of the filter.) Now use the Kuroda identities to obtain the configuration shown in Figure 8.22,

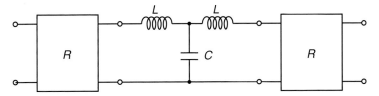

Figure 8.21   Filter with transmission line extensions.

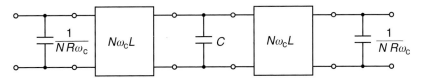

Figure 8.22   Filter after application of Kuroda identities.

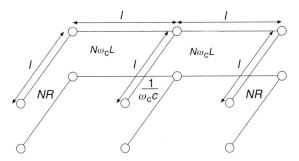

Figure 8.23   Transmission line realisation.

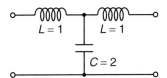

Figure 8.24   Prototype lumped component filter.

where $N = 1 + (\omega_c L/R)$. In transmission line terms, we now have the configuration shown in Figure 8.23 with $l = \lambda/8$ at frequency ω_c. This configuration only involves open-circuit stubs and is easily implemented in microstrip technology.

Example   Design a low-pass microstrip filter, based on a three-component discrete filter, with 3 dB point at 1 GHz and 50 Ω terminations.

Consider the lumped component prototype filter (1 Ω impedance and 3 dB bandwidth of 1 rad/s) shown in Figure 8.24. In terms of prototype values, this will translate to the transmission line filter shown in Figure 8.25. Rescaling for 50 Ω terminations we obtain the configuration of Figure 8.26. Table 8.6 shows the performance of the filter in terms

Table 8.6 *Performance of a stripline filter*

Frequency (MHz)	Relative power	Frequency (MHz)	Relative power	Frequency (MHz)	Relative power
0	1.000	3100	0.720	6800	0.128
500	0.995	3200	0.872	6900	0.280
800	0.872	3500	0.995	7000	0.500
900	0.720	4000	1.000	7100	0.720
1000	0.500	4500	0.995	7200	0.872
1100	0.280	4800	0.872	7500	0.995
1200	0.128	4900	0.720	8000	1.000
1500	0.005	5000	0.500	8500	0.995
2000	0.000	5100	0.280	8800	0.872
2500	0.005	5200	0.128	8900	0.720
2800	0.128	5500	0.005	9000	0.500
2900	0.280	6000	0.000	9100	0.280
3000	0.500	6500	0.005	9200	0.128

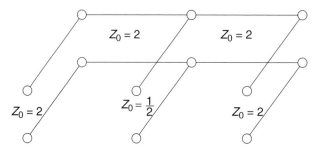

Figure 8.25 Prototype transmission line filter.

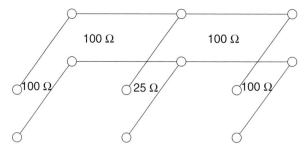

Figure 8.26 Final transmission line filter.

of transducer gain. It will be noted that the desired low-pass behaviour is exhibited for frequencies below 2000 MHz but above this frequency the filter also has significant response. As mentioned earlier, the mapping from discrete components to transmission line stubs is not perfect. Up to the frequency $2\omega_c$ the correspondence is sufficient, but

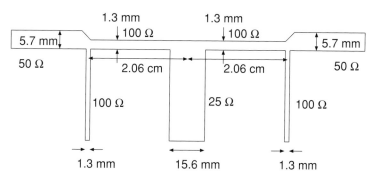

Figure 8.27   Microstrip realisation of the filter.

Figure 8.28   Simple parallel microstrip filter.

above this frequency the periodic variation of transmission line properties will mean that the filter response is also periodic. The result is additional band-pass responses from 3000 MHz to 5000 MHz, 7000 MHz to 9000 MHz, and so on. Obviously, if a lumped component filter is to be converted to microstrip form, the impact of the additional response should be carefully considered.

Consider a microstrip realisation of the above filter using an FR4 printed circuit board with 3 mm thick substrate ($\epsilon_r = 4.4$). We will have $\epsilon_{\text{eff}} = 3.33$ and so a wavelength on the microstrip (at 1 GHz) will be 16.5 cm and hence $\lambda/8$ will be 2.06 cm. From the microstrip design formulae of Chapter 4, a 100 Ω line will require a width of 1.3 mm, a 50 Ω line a width of 5.7 mm and a 25 Ω line a width of 15.6 mm. The final microstrip filter is shown in Figure 8.27.

Due to end effects, the open stubs will need to be slightly shorter than the 2.06 cm calculated above. A length

$$\Delta = 0.42 \frac{\epsilon_{\text{eff}} + 0.3}{\epsilon_{\text{eff}} - 0.258} \frac{w + 0.264h}{w + 0.8h} h \tag{8.17}$$

will need to be removed from their open ends (h is the substrate thickness and w is the microstrip width).

A band-pass filter can be formed from two closely spaced parallel microstrip lines, each line open circuit at one end (see Figure 8.28). For quarter-wave lines, this filter can be regarded as the microstrip directional coupler of Section 6.9 with two of its ports left open circuit. From the S matrix of the coupler, we find that $S_{12} = -2jc_{\lambda/4}\sqrt{1 - c_{\lambda/4}^2}$ for the two-port network of Figure 8.28 (note that $c_{\lambda/4} = |(Z_e - Z_o)/(Z_e + Z_o)|$, where Z_e

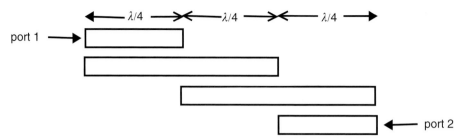

Figure 8.29   Parallel microstrip filter.

and Z_o are the even and odd mode characteristic impedances of the parallel transmission lines). Consequently, at a frequency for which the transmission line length l is a quarter wavelength, the transducer gain is given by $G_T = 4c_{\lambda/4}^2(1 - c_{\lambda/4}^2)$ when the ports are terminated in the characteristic impedance $Z_0 = \sqrt{Z_e Z_o}$. There will be maximum gain when $c_{\lambda/4} = 1/\sqrt{2}$. If the length l of the transmission lines is different from a quarter wavelength, the coupling coefficient will be given by

$$c_l = \frac{jc_{\lambda/4} \sin \frac{2\pi l}{\lambda}}{\sqrt{1 - c_{\lambda/4}^2} \cos \frac{2\pi l}{\lambda} + j \sin \frac{2\pi l}{\lambda}} \tag{8.18}$$

and from which it will be noted that the gain falls away either side of the frequency for which l is a quarter wavelength. Although the above filter will only exhibit modest performance, numerous copies can be cascaded (see Figure 8.29) to provide a filter with extremely good characteristics. In a realistic design, however, the microstrip spacings and widths will be adjusted for optimum performance.

EXERCISES

(1) Design a three-element low-pass Butterworth filter with a 3 dB frequency of 5 MHz and 50 Ω terminations.

(2) Design a five-element low-pass Chebyshev filter (1 dB passband ripple) with a bandwidth of 20 MHz and 50 Ω terminations.

(3) Design a low-pass Butterworth filter with a 3 dB frequency of 10 MHz and 50 Ω terminations. Choose sufficient elements such that there is at least 30 dB of attenuation at 20 MHz.

(4) Design a three-element band-pass Butterworth filter with a centre frequency of 100 MHz, a 3 dB bandwidth of 10 MHz and 100 Ω terminations.

(5) Design a three-element low-pass Butterworth filter with a 3 dB frequency of 1.5 GHz and 50 Ω terminations. Convert this filter into microstrip form.

(6) By considering a directional coupler with two of its ports open circuit, use the S matrix of the coupler to find an expression for the S matrix of a parallel microstrip filter. (Note that the incoming and outgoing voltages at an open-circuit port are of equal magnitude.)

SOURCES

Bowick, C. 1982. *RF Circuit Design*. Boston, MA: Newnes.

Gupta, K. C., Garg, R. and Bahl, I. J. 1979. *Microstrip Lines and Slot Lines*. Dedham, MA: Artech House.

Hayward, W. 1994. *Introduction to Radio Frequency Design*. Newark, CT: American Radio Relay League.

Ludwig, R. and Bretchko, P. 1990. *RF Circuit Design: Theory and Application*. Upper Saddle River, NJ: Prentice-Hall.

Pozar, D. M. 1998. *Microwave Engineering* (2nd edn). New York: John Wiley.

Pozar, D. M. 2001. *Microwave and RF Design of Wireless Systems*. New York: John Wiley.

Straw, R. Dean (ed.). 1999. *The ARRL Handbook* (77th edn). Newark, CT: American Radio Relay League.

9 Electromagnetic waves

The work of Maxwell, and other pioneers, led to the development of equations that fully describe electromagnetic phenomena. These equations reveal an intimate connection between electric and magnetic fields and show that there are circumstances in which one field cannot exist without the other. In particular, the equations predict the existence of electromagnetic waves and demonstrate that, for a given electric wave field, there is an associated magnetic field. In 1887, Hertz verified the existence of electromagnetic waves and also the property that the wave amplitude decays as the inverse of distance from the source. This slow decay in amplitude is the property that makes electromagnetic waves an effective means of communication, as demonstrated by Marconi at the end of the nineteenth century. Unfortunately, electromagnetic waves have a much more complex structure than many other waves (sound waves, for example) and this complexity must be understood if they are to be successfully exploited. As a consequence, the current chapter explores the basic theory of electromagnetic waves.

9.1 Maxwell's equations

The equations of Maxwell take the form

$$\nabla \cdot \mathcal{B} = 0 \tag{9.1}$$

$$\nabla \cdot \mathcal{D} = \rho \tag{9.2}$$

$$\nabla \times \mathcal{E} = -\frac{\partial \mathcal{B}}{\partial t} \tag{9.3}$$

$$\nabla \times \mathcal{H} = \frac{\partial \mathcal{D}}{\partial t} + \mathcal{J}, \tag{9.4}$$

where $\mathcal{E}$, $\mathcal{D}$, $\mathcal{B}$, $\mathcal{H}$ are the electric intensity (volts per metre), electric flux density (coulombs per square metre), magnetic flux density (webers per square metre) and magnetic intensity (amperes per metre), respectively. The fields have, as their sources, the electric charge distribution ρ and current distribution $\mathcal{J}$ ($\mathcal{J} = \rho v$, where v is the velocity field of the charge). These sources have units of coulombs per cubic metre and amperes per square metre, respectively. For a region of space that is filled with isotropic dielectric material,

$$B = \mu \mathcal{H} \tag{9.5}$$

$$D = \epsilon \mathcal{E}, \tag{9.6}$$

where μ and ϵ are permeability and permittivity of the material. If ϵ_0 is the permittivity of free space (8.854×10^{-12} F/m), we can define the relative permittivity ϵ_r of the material by $\epsilon = \epsilon_r \epsilon_0$. Likewise, if μ_0 is the permeability of free space ($4\pi \times 10^{-7}$ H/m), we can define the relative permeability μ_r of the material by $\mu = \mu_r \mu_0$.

An electromagnetic field manifests itself through its effect on charge. Notably, an isolated charge q will experience the *Lorentz force*

$$\mathcal{F} = q(\mathcal{E} + v \times B) \tag{9.7}$$

when travelling with velocity v. More often than not, however, we are more interested in charges flowing through a *conducting material*. In this case, the motion of a charge is quite complex since it is heavily constrained by the surrounding material. Fortunately, in most situations, the charge motion is well modelled by the generalised *Ohm's Law*. This states that an electric field $\mathcal{E}$ will cause a current density $\mathcal{J} = \sigma \mathcal{E}$, where σ (S/m) is a material constant known as the *conductivity*.

The Maxwell equations do not provide a unique description of the electromagnetic field unless they are supplemented by suitable boundary conditions. If a solution to the Maxwell equations satisfies the requisite boundary conditions, it can be proven that it is the one and only such solution. Consequently, the solution process will often consist of searching for a linear combination of simple solutions that satisfies the appropriate boundary conditions. On a boundary between two dissimilar materials of finite conductivity, the boundary conditions require that $\mathcal{H} \times n$ (tangential components of $\mathcal{H}$), $B \cdot n$ (normal component of B), $\mathcal{E} \times n$ (tangential components of $\mathcal{E}$) and $D \cdot n$ (normal component of D) are all continuous across the boundary (n is a unit vector that is normal to the boundary). An important special case occurs when one of the materials is a perfect conductor ($\sigma = \infty$) which is a good approximation for metallic materials. In this case, the charge will accumulate at the conductor surface to form a surface charge density ρ_s (Coulombs per square metre) and surface current density $\mathcal{J}_s$ (A/m). Inside the conductor the electric field will be zero with $n \times \mathcal{E} = 0$ on the surface (i.e., the tangential electric field is zero). In addition, $n \times \mathcal{H} = \mathcal{J}_s$, $B \cdot n = 0$ and $D \cdot n = \rho_s$ on the surface of the conductor. It should be noted, however, that both surface current and charge will adjust themselves such that the conditions involving these quantities are automatically satisfied.

9.2　Power flow

From Maxwell's equations, it can be shown that the power $\mathcal{P}_S$ delivered by the field sources within a volume V is related to the resulting fields through

$$\mathcal{P}_S = \int \sigma \mathcal{E} \cdot \mathcal{E} \, dV + \frac{\partial}{\partial t} \int_V \left(\frac{\epsilon \mathcal{E} \cdot \mathcal{E}}{2} + \frac{\mathcal{H} \cdot \mathcal{H}}{2\mu} \right) dV + \oint_S (\mathcal{E} \times \mathcal{H}) \cdot d\mathcal{S} \qquad (9.8)$$

where S is the surface of volume V. The first term on the right represents *ohmic loss* (loss through heating effects) within V, the second term represents the rate of change of field energy W ($= \int_V [\epsilon \mathcal{E} \cdot \mathcal{E}/2 + \mathcal{H} \cdot \mathcal{H}/2\mu] \, dV$) and the last term represents the flow of field energy out of the volume. The vector

$$\mathcal{P} = \mathcal{E} \times \mathcal{H} \qquad (9.9)$$

is known as Poynting's vector and is the instantaneous rate of energy flow across a unit area that is perpendicular to $\mathcal{P}$.

9.3　Electromagnetic waves

As mentioned earlier, the effect of changes in the electromagnetic field sources will travel outwards in the manner of a wave. This can be seen in general from the Maxwell equations

$$\nabla \times \mathcal{E} = -\mu \frac{\partial \mathcal{H}}{\partial t} \qquad (9.10)$$

$$\nabla \cdot \mathcal{E} = 0 \qquad (9.11)$$

$$\nabla \times \mathcal{H} = \epsilon \frac{\partial \mathcal{E}}{\partial t} \qquad (9.12)$$

$$\nabla \cdot \mathcal{H} = 0 \qquad (9.13)$$

which are valid outside the sources. (Note that we have assumed a propagation medium such that μ and ϵ are constant.) From the curl Maxwell equations, we have

$$\nabla \times \nabla \times \mathcal{E} = -\mu \frac{\partial}{\partial t} \nabla \times \mathcal{H}$$

$$= -\mu\epsilon \frac{\partial^2 \mathcal{E}}{\partial t^2} \qquad (9.14)$$

and, by the vector identity $\nabla \times (\nabla \times \mathcal{G}) = \nabla(\nabla \cdot \mathcal{G}) - \nabla^2 \mathcal{G}$,

$$\nabla(\nabla \cdot \mathcal{E}) - \nabla^2 \mathcal{E} = -\mu\epsilon \frac{\partial^2 \mathcal{E}}{\partial t^2}. \qquad (9.15)$$

Since $\nabla \cdot \mathcal{E} = 0$, the above equation reduces to

$$\nabla^2 \mathcal{E} - \frac{1}{c^2} \frac{\partial^2 \mathcal{E}}{\partial t^2} = 0, \tag{9.16}$$

where $c = (\mu \epsilon)^{-1/2}$. In a similar fashion,

$$\nabla^2 \mathcal{H} - \frac{1}{c^2} \frac{\partial^2 \mathcal{H}}{\partial t^2} = 0. \tag{9.17}$$

Both $\mathcal{E}$ and $\mathcal{H}$ satisfy wave equations with propagation speed c ($c_0 = 3 \times 10^8$ m/s in free space), but they will also need to satisfy the original Maxwell equations and these will impose additional constraints.

The wave equations will exhibit plane wave solutions (solutions that are constant in any plane perpendicular to the propagation direction) and these will take the form

$$\mathcal{E} = \mathcal{E}\left(t - \frac{\mathbf{r} \cdot \hat{\mathbf{p}}}{c}\right) \tag{9.18}$$

$$\mathcal{H} = \mathcal{H}\left(t - \frac{\mathbf{r} \cdot \hat{\mathbf{p}}}{c}\right), \tag{9.19}$$

where $\hat{\mathbf{p}}$ is a unit vector in the propagation direction. The Maxwell curl equations will imply

$$-\frac{1}{c}\hat{\mathbf{p}} \times \frac{\partial \mathcal{E}}{\partial t} = -\mu \frac{\partial \mathcal{H}}{\partial t} \tag{9.20}$$

and

$$-\frac{1}{c}\hat{\mathbf{p}} \times \frac{\partial \mathcal{H}}{\partial t} = \epsilon \frac{\partial \mathcal{E}}{\partial t}. \tag{9.21}$$

On integrating these expressions, we obtain

$$\mathcal{H} = \frac{\hat{\mathbf{p}} \times \mathcal{E}}{\mu c} \tag{9.22}$$

and

$$\mathcal{E} = -\frac{\hat{\mathbf{p}} \times \mathcal{H}}{\epsilon c}. \tag{9.23}$$

In addition, the divergence equations imply

$$\hat{\mathbf{p}} \cdot \mathcal{E} = \hat{\mathbf{p}} \cdot \mathcal{H} = 0. \tag{9.24}$$

Consequently, $\mathcal{E}$ can be an arbitrary vector function of $t - (\mathbf{r} \cdot \hat{\mathbf{p}}/c)$ providing the propagation direction $\hat{\mathbf{p}}$ is orthogonal to $\mathcal{E}$. The magnetic field, however, must take the form

$$\mathcal{H} = \frac{\hat{\mathbf{p}} \times \mathcal{E}}{\eta}, \tag{9.25}$$

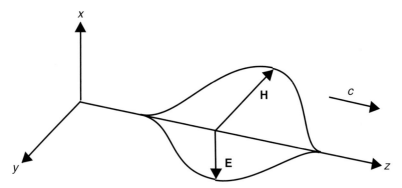

Figure 9.1   A plane electromagnetic pulse.

where $\eta = \sqrt{\mu/\epsilon}$ is the impedance of the propagation medium (note that the impedance of free space η_0 is approximately 377 Ω). From Equations 9.9, 9.24 and 9.25 we find the Poynting vector to be of the form $(\hat{\mathbf{p}}/\eta)\mathcal{E} \cdot \mathcal{E}$ and so it is clear that electromagnetic waves will transport energy in the propagation direction (Figure 9.1).

In radio systems, we are normally concerned with signals that have a relatively narrow bandwidth and so a single frequency assumption is often sufficient for the analysis of electromagnetic aspects. In terms of the fields, this *time-harmonic* assumption is expressed as

$$\mathcal{H}(\mathbf{r}, t) = \Re\{\mathbf{H}(\mathbf{r})\, e^{j\omega t}\} \tag{9.26}$$

$$\mathcal{E}(\mathbf{r}, t) = \Re\{\mathbf{E}(\mathbf{r})\, e^{j\omega t}\} \tag{9.27}$$

(note that non-italic capitals are used to represent the fields with the time dependence factored out). Fields $\mathbf{H}(\mathbf{r})$ and $\mathbf{E}(\mathbf{r})$ will satisfy the time-harmonic Maxwell equations

$$\nabla \times \mathbf{E} = -j\omega\mu\mathbf{H} \tag{9.28}$$

and

$$\nabla \times \mathbf{H} = j\omega\epsilon\mathbf{E} + \mathbf{J}, \tag{9.29}$$

where $\mathcal{J} = \Re\{\mathbf{J}\, e^{j\omega t}\}$. The divergence Maxwell equations will be automatically satisfied as long as the charge density satisfies the time-harmonic continuity equation $j\omega\rho + \nabla \cdot \mathbf{J} = 0$. A time-harmonic plane wave in free space will have the form

$$\mathcal{E} = \Re\left\{\mathbf{E}_0\, e^{j\omega(t - \frac{\hat{\mathbf{p}} \cdot \mathbf{r}}{c})}\right\} \tag{9.30}$$

$$\mathcal{H} = \Re\left\{\mathbf{H}_0\, e^{j\omega(t - \frac{\hat{\mathbf{p}} \cdot \mathbf{r}}{c})}\right\}, \tag{9.31}$$

where $\hat{\mathbf{p}}$ is a unit vector in the direction of propagation. Maxwell's equations imply that

$$\mathbf{H}_0 = \frac{\hat{\mathbf{p}} \times \mathbf{E}_0}{\eta} \tag{9.32}$$

together with the relation $\hat{\mathbf{p}} \cdot \mathbf{E}_0 = 0$. It will be noted that both electric and magnetic fields are orthogonal to the direction of propagation and, because of this, the plane wave is known as a TEM (transverse electromagnetic) wave. In the time-harmonic case, we find it more convenient to specify the energy flow in terms of the average flow over a wave period. This leads to the time-harmonic Poynting vector

$$\mathbf{P} = \frac{1}{2}\Re\{\mathbf{E} \times \mathbf{H}\} \tag{9.33}$$

which represents the average rate of flow of energy across a unit area that is perpendicular to the direction of the vector. In the case of a plane-harmonic wave

$$\mathbf{P} = \frac{\hat{\mathbf{p}}}{2\eta}\mathbf{E} \cdot \mathbf{E}^* \tag{9.34}$$

where $\hat{\mathbf{p}}$ is the propagation direction.

If the direction of propagation is the z-axis ($\hat{\mathbf{p}} = \hat{\mathbf{z}}$), the electric field has the form

$$\mathbf{E} = \mathbf{E}_0 \, e^{-j\beta z}, \tag{9.35}$$

where $\beta = \omega/c$ is the propagation constant ($\beta = 2\pi/\lambda$ in terms of wavelength λ) and $\mathbf{E}_0$ has no component in the $\hat{\mathbf{z}}$ direction. It will be noted that, up to an arbitrary phase, $\mathbf{E}$ can be expressed as

$$\mathbf{E} = E_0(\cos\gamma\hat{\mathbf{x}} + \sin\gamma\, e^{j\delta}\hat{\mathbf{y}})\, e^{-j\beta z}. \tag{9.36}$$

For $\delta = 0$, the $\hat{\mathbf{x}}$ and $\hat{\mathbf{y}}$ components will be in phase and we will have *linear polarisation*. That is, the electric field $\Re(\mathbf{E}\,e^{j\omega t})$ will always be proportional to a constant vector at angle γ to the $\hat{\mathbf{x}}$ direction. If $\delta \neq 0$, the direction of the electric field will rotate about the propagation direction. Two important special cases are when $\gamma = 45°$ and either $\delta = 90°$ or $\delta = -90°$. In these cases,

$$\mathcal{E}_{\pm}(\mathbf{r}, t) = \frac{E_0}{\sqrt{2}}\left\{\cos\left[\omega\left(t - \frac{z}{c}\right)\right]\hat{\mathbf{x}} \mp \sin\left[\omega\left(t - \frac{z}{c}\right)\right]\hat{\mathbf{y}}\right\} \tag{9.37}$$

from which it will be noted that the electric field traces out a circle about the propagation direction. If the circle is traced out in an anticlockwise fashion (the wave travelling away from the observer) we have left-hand circular polarisation, otherwise we have right-hand circular polarisation. It should be noted that any linear polarisation can be expressed as a combination of left- and right-hand circular polarisations. A simple example of this is given by

$$\mathbf{E} = E_0\hat{\mathbf{x}}\, e^{-j\beta z} \tag{9.38}$$

$$= \mathbf{E}_+ + \mathbf{E}_- \tag{9.39}$$

$$= \frac{E_0}{2}(\hat{\mathbf{x}} + j\hat{\mathbf{y}})\, e^{-j\beta z} + \frac{E_0}{2}(\hat{\mathbf{x}} - j\hat{\mathbf{y}})\, e^{-j\beta z} \tag{9.40}$$

and is illustrated in Figure 9.2. Such decompositions are important since there are certain media (the Earth's ionosphere is an example) for which the right and left circular polarisations have different propagation speeds. Consequently, after propagation

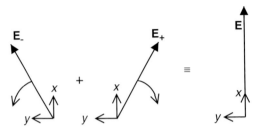

Figure 9.2 The decomposition of linear polarisation into circularly polarised modes.

through such a medium, a linear polarisation will have its orientation changed. For a medium such as the ionosphere, this reorientation can be quite unpredictable and so has important consequences for the planning of communication links involving satellites. To overcome this problem, it is common to use an antenna that only responds to one of the two possible circular polarised modes.

If a medium has finite conductivity σ, electric field $\mathbf{E}$ will cause a current $\mathbf{J} = \sigma\mathbf{E}$ to flow. As a consequence, the fields will satisfy time-harmonic Maxwell equations of the form

$$\nabla \times \mathbf{E} = -j\omega\mu\mathbf{H} \tag{9.41}$$

and

$$\nabla \times \mathbf{H} = j\omega\epsilon_{\text{eff}}\mathbf{E}, \tag{9.42}$$

where $\epsilon_{\text{eff}} = \epsilon - (j\sigma/\omega)$ is known as the *effective permittivity* of the medium. For a plane wave, both the impedance and propagation constant will now be complex

$$\eta = \sqrt{\frac{\mu}{\epsilon_{\text{eff}}}} = \sqrt{\frac{\mu}{\epsilon}} \frac{1}{\sqrt{1 - \frac{j\sigma}{\omega\epsilon}}} \tag{9.43}$$

and

$$\beta = \omega\sqrt{\mu\epsilon_{\text{eff}}} = \frac{\omega}{c}\sqrt{1 - \frac{j\sigma}{\omega\epsilon}}. \tag{9.44}$$

The imaginary part of β will give rise to an exponential decay of wave amplitude in the direction of propagation (attenuation). For a wave travelling in the z-direction

$$\mathcal{E} = \exp(\Im\{\beta\}z)\,\Re\{\mathbf{E}_0\exp j(\omega t - \Re\{\beta\}z)\} \tag{9.45}$$

and

$$\mathcal{H} = \exp(\Im\{\beta\}z)\,\Re\left\{\frac{\hat{\mathbf{z}} \times \mathbf{E}_0}{\eta}\exp j(\omega t - \Re\{\beta\}z)\right\}, \tag{9.46}$$

where $\mathbf{E}_0$ is the vector amplitude of the electric field when $z = 0$. Since the real part of β is frequency dependent, the speed of the phase front $V_{\text{p}} = \omega/\Re\{\beta\}$ will also be frequency dependent. A medium that gives rise to a frequency dependent phase speed is

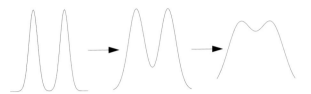

Figure 9.3 Dispersion of pulses.

said to be *dispersive* and such media can have serious consequences for communication systems. A modulated signal will have components over a range of frequencies and these can propagate at different speeds with the result that the modulation can lose its integrity after a sufficient propagation distance. In the case of a pulse modulated signal, the pulses will spread out during propagation and eventually merge (see Figure 9.3). At this point, the information that is represented by the modulation will be lost. As a consequence, dispersion effects place a fundamental limitation upon communication systems and must be taken into account in system design.

Because a pulse will consist of components over a range of frequencies, it becomes difficult to give meaning to the phase speed when the medium is dispersive. In this case, a more useful concept is that of group speed (the speed at which the energy of the pulse propagates). This can usually be interpreted as the speed of the pulse peak. Consider the case where there are only two components with frequencies separated by $\delta\omega$,

$$\mathcal{E}_1 = E_0 \cos(\beta z - \omega t)\hat{\mathbf{x}} \tag{9.47}$$

and

$$\mathcal{E}_2 = E_0 \cos[(\beta + \delta\beta)z - (\omega + \delta\omega)t]\hat{\mathbf{x}} \tag{9.48}$$

from which the total field is

$$\mathcal{E} = \mathcal{E}_1 + \mathcal{E}_2 = 2E_0 \cos\left(\frac{\delta\beta z - \delta\omega t}{2}\right) \cos\left[\left(\beta + \frac{\delta\beta}{2}\right)z - \left(\omega + \frac{\delta\omega}{2}\right)t\right]\hat{\mathbf{x}}. \tag{9.49}$$

Let V_g be the speed at which the modulation peak travels, then $V_g = \delta\omega/\delta\beta$. In general, the group speed (Figure 9.4) is given by

$$v_g = \frac{\partial\omega}{\partial\beta}. \tag{9.50}$$

Thus far, we have only considered propagation through a homogeneous space. Realistic situations, however, will often involve a complex combination of different materials and it is important to understand the effect of such complexity upon propagation. The simplest example is a plane wave that is normally incident upon a plane interface between two different media (Figure 9.5). At the interface, some of the energy will be

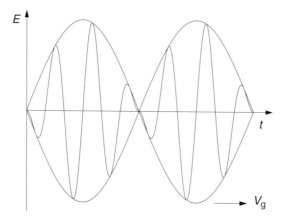

Figure 9.4   Group speed for a simple two-component signal.

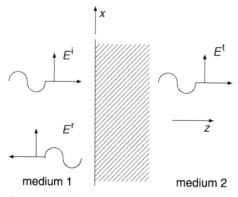

Figure 9.5   Reflection and transmission at a plane interface.

reflected and some transmitted. Let the incident wave have the form

$$\mathbf{E}^i(z) = E_0^i\, e^{-j\beta_1 z}\hat{\mathbf{x}}$$ (9.51)

$$\mathbf{H}^i(z) = H_0^i\, e^{-j\beta_1 z}\hat{\mathbf{y}},$$ (9.52)

the transmitted wave the form

$$\mathbf{E}^t(z) = E_0^t\, e^{-j\beta_2 z}\hat{\mathbf{x}}$$ (9.53)

$$\mathbf{H}^t(z) = H_0^t\, e^{-j\beta_2 z}\hat{\mathbf{y}},$$ (9.54)

and the reflected wave the form

$$\mathbf{E}^r(z) = E_0^r\, e^{+j\beta_1 z}\hat{\mathbf{x}}$$ (9.55)

$$\mathbf{H}^r(z) = H_0^r\, e^{+j\beta_1 z}\hat{\mathbf{y}},$$ (9.56)

where β_1 and β_2 are the propagation constants in media 1 and 2, respectively. Since we are dealing with plane waves, the electric and magnetic fields for these waves are related through

$$E_0^i = \eta_1 H_0^i \tag{9.57}$$

$$E_0^r = -\eta_1 H_0^r \tag{9.58}$$

$$E_0^t = \eta_2 H_0^t, \tag{9.59}$$

where η_1 and η_2 are the impedances of media 1 and 2, respectively. Boundary conditions require the continuity of tangential components ($\hat{x}$ and $\hat{y}$ components) of both $\mathbf{H}$ and $\mathbf{E}$ at the interface. Consequently,

$$E_0^i + E_0^r = E_0^t \tag{9.60}$$

$$H_0^i + H_0^r = H_0^t. \tag{9.61}$$

The last of these conditions can be replaced by

$$\frac{E_0^i}{\eta_1} - \frac{E_0^r}{\eta_1} = \frac{E_0^t}{\eta_2} \tag{9.62}$$

and so, eliminating between the above equations, we obtain the magnitudes of the transmitted and reflected fields in terms of the incident field

$$E_0^t = \frac{2\eta_2}{\eta_1 + \eta_2} E_0^i \tag{9.63}$$

$$E_0^r = \frac{\eta_2 - \eta_1}{\eta_1 + \eta_2} E_0^i. \tag{9.64}$$

The above results can be used to estimate the attenuation of a wave in passing through a strongly conducting screen (see Figure 9.6). Medium 2 (impedance η and propagation constant β) will be strongly attenuating and mediums 1 and 3 are assumed to consist of free space (impedance η_o). The strong attenuation in medium 2 means that we can ignore E^- (the reflection at the second interface) and hence treat the interfaces separately. At

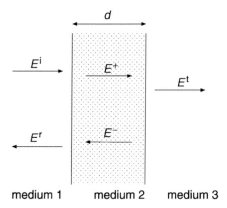

medium 1　　medium 2　　medium 3

Figure 9.6　Transmission through a screen (arrows indicate propagation directions).

the first interface the transmitted field E^+ will have the value

$$E_0^+ = \frac{2\eta}{\eta_0 + \eta} E_0^i.$$ (9.65)

Field E^+ will suffer attenuation $\exp(-|\Im\{\beta\}|d)$ in passage through the screen and hence have magnitude $E_0^+ \exp(-|\Im\{\beta\}|d)$ at the second interface. At this interface, the transmitted field will have magnitude

$$E^t = \frac{2\eta_0}{\eta_0 + \eta} E_0^+ e^{-|\Im\{\beta\}|d}$$

$$= \frac{4\eta_0\eta}{(\eta_0 + \eta)^2} E_0^i e^{-|\Im\{\beta\}|d}.$$ (9.66)

Since the screen is strongly conducting, we can use the approximations

$$\beta \approx (1 - j)\sqrt{\frac{\omega\mu\sigma}{2}}$$ (9.67)

$$\eta \approx (1 + j)\sqrt{\frac{\omega\mu}{2\sigma}},$$ (9.68)

where σ is the conductivity of the screen and μ is its permeability. Consequently, since $\eta \ll \eta_0$ the original wave will be reduced in amplitude by factor $4|\eta/\eta_0| \exp(-\sqrt{\omega\mu\sigma/2}d)$ in passage through the screen.

Example An RF system needs to be screened by an aluminium box that provides 120 dB of attenuation at 10 MHz. Calculate the thickness of metal that is required.

For aluminium, $\sigma = 3.6 \times 10^7$ S/m and we take $\mu = \mu_0$. We will need the thickness d to satisfy $4|\eta/\eta_0| \exp(-\sqrt{\omega\mu\sigma/2}d) = 10^{-6}$ (the ratio of amplitudes for 120 dB of attenuation) with $\mu = 4\pi \times 10^{-7}$ H/m and $\omega = 2\pi \times 10^7$ rad/s. Note that $|\eta/\eta_0| = \sqrt{\omega\mu/\sigma}/120\pi = 3.93 \times 10^{-6}$ and $\sqrt{\omega\mu\sigma/2} = 3.76 \times 10^4$. Consequently, we will need $\exp(-3.76 \times 10^4 d) = 0.0636$ from which $d = 0.0733$ mm.

Another consequence of wave attenuation is the confinement of RF currents to regions near the surface of a metal. The current, and hence the electric field, will be tangential to the metal surface with the wave propagating into the metal (see Figure 9.7). From previous considerations, it is clear that the electric field (and hence the current density) will decay exponentially with the distance into the metal

$$|J| = \sigma|E^t| = \sigma|E_0^t| \exp\left(-\frac{z}{\delta}\right),$$ (9.69)

where $\delta = \sqrt{2/\omega\mu\sigma}$ is known as the *skin depth* and E_0^t is the tangential electric field at the surface of the conductor. For a metal surface, δ will be very small at high frequencies (the *skin effect*) and there will be an effective surface current distribution $\mathbf{J}_s$ that satisfies

$$\mathbf{J}_s = \frac{1}{R_s}\mathbf{E}_s,$$ (9.70)

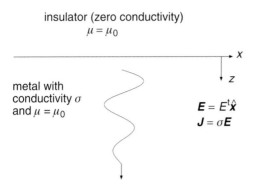

insulator (zero conductivity)
$\mu = \mu_0$

metal with
conductivity σ
and $\mu = \mu_0$

$E = E^t \hat{x}$
$J = \sigma E$

Figure 9.7   Attenuation in a current carrying conductor.

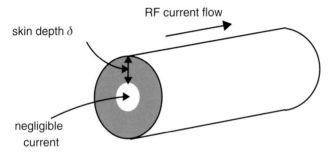

RF current flow

skin depth δ

negligible
current

Figure 9.8   The skin effect on a wire conductor.

where R_s is the surface resistance ($R_s = 1/\delta\sigma = \sqrt{\omega\mu/2\sigma}$) and $\mathbf{E}_s$ is the electric field at the surface. The surface resistance will result in a loss of power through ohmic heating ($P_{\text{ohm}} = \frac{1}{2}\int R_s|\mathbf{J}_s|^2\,dS$ with the integral performed over the metal surface) and this will increase with frequency. Since most RF components rely on good conductivity for their effective operation, the skin effect will have a severe impact upon component performance at high frequencies. For example (Figure 9.8), a circular wire of radius a will have resistance $1/2\delta\sigma\pi a$ per unit length and, since δ decreases with frequency, the resistance will rise. If an inductor is wound from such wire, the inductor resistance will rise with frequency and, as a consequence, there will be a reduction in the Q (Q is inversely proportional to the resistance).

9.4   Oblique incidence

Although reflections at normal incidence constitute an important case, there are many situations in which propagation via oblique reflections is the dominant mechanism (Figure 9.9). We first consider the case of a plane wave that is obliquely incident upon a perfectly conducting surface. For time-harmonic fields, all components will be zero inside the conductor and the electric field will satisfy the condition

$$\mathbf{E} \times \mathbf{n} = 0 \tag{9.71}$$

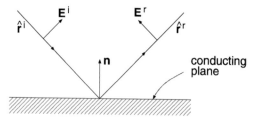

Figure 9.9   Oblique incidence of plane waves.

on the surface of the conductor (n is the unit normal to the surface). The condition on the tangential $\mathbf{E}$ field is all that is required to establish a unique solution to Maxwell's equations (the surface current will automatically adjust itself to satisfy $\mathbf{H} \times n = \mathbf{J_s}$). Given an incident plane wave field $(\mathbf{H^i}, \mathbf{E^i})$ with unit vector $\hat{\mathbf{r}}^i$ in the propagation direction, we would like to determine the reflected wave field $(\mathbf{H^r}, \mathbf{E^r})$ and the unit vector $\hat{\mathbf{r}}^r$ in its propagation direction. (It should be noted that we only need to calculate the reflected electric field since $\mathbf{H^r} = \hat{\mathbf{r}}^r \times \mathbf{E^r}/\eta$, where η is the impedance of the medium to the fore of the reflecting surface). The incident and reflected electric fields will take the form

$$\mathbf{E^i} = \mathbf{E}_0^i \exp[-\beta\, \hat{\mathbf{r}}^i \cdot (\mathbf{r} - \mathbf{r}_0)] \tag{9.72}$$

$$\mathbf{E^r} = \mathbf{E}_0^r \exp[-\beta\, \hat{\mathbf{r}}^r \cdot (\mathbf{r} - \mathbf{r}_0)], \tag{9.73}$$

where $\mathbf{E}_0^r$ and $\mathbf{E}_0^i$ are constant vectors perpendicular to the propagation directions $\hat{\mathbf{r}}^r$ and $\hat{\mathbf{r}}^i$, respectively. Since we require $n \times \mathbf{E} = 0$ on the reflecting surface, we will have

$$n \times \left\{ \mathbf{E}_0^i \exp[-\beta\, \hat{\mathbf{r}}^i \cdot (\mathbf{r} - \mathbf{r}_0)] + \mathbf{E}_0^r \exp[-\beta\, \hat{\mathbf{r}}^r \cdot (\mathbf{r} - \mathbf{r}_0)] \right\} = 0 \tag{9.74}$$

for points satisfying $(\mathbf{r} - \mathbf{r}_0) \cdot n = 0$ ($\mathbf{r}_0$ is a reference point on the reflecting surface). The above equations will imply that

$$n \times (\mathbf{E}_0^i + \mathbf{E}_0^r) = 0 \tag{9.75}$$

and

$$\hat{\mathbf{r}}^i \cdot (\mathbf{r} - \mathbf{r}_0) = \hat{\mathbf{r}}^r \cdot (\mathbf{r} - \mathbf{r}_0) \tag{9.76}$$

for points $\mathbf{r}$ on the surface (those satisfying $(\mathbf{r} - \mathbf{r}_0) \cdot n = 0$). Condition 9.76 will be satisfied if

$$\hat{\mathbf{r}}^r = \hat{\mathbf{r}}^i - 2n(n \cdot \hat{\mathbf{r}}^i) \tag{9.77}$$

and Condition 9.75 will be satisfied if

$$\mathbf{E}_0^r = n[n \cdot (\mathbf{E}_0^i + \mathbf{E}_0^r)] - \mathbf{E}_0^i, \tag{9.78}$$

Expression 9.78 only determines the tangential components of $\mathbf{E}_0^r$ in terms of $\mathbf{E}_0^i$ and not the normal component. It should be noted, however, that the reflected field will need to be orthogonal to the propagation direction $\hat{\mathbf{r}}^r$ (i.e., $\hat{\mathbf{r}}^r \cdot \mathbf{E}_0^r = 0$). This condition, together with Conditions 9.77 and 9.78, implies that $\mathbf{n} \cdot \mathbf{E}_0^i = \mathbf{n} \cdot \mathbf{E}_0^r$ and, as a consequence,

$$\mathbf{E}_0^r = 2\mathbf{n}(\mathbf{n} \cdot \mathbf{E}_0^i) - \mathbf{E}_0^i. \tag{9.79}$$

We are often concerned with reflections from imperfect conductors and, for such cases, the above expression will generalise to

$$\mathbf{E}_0^r = -R_H \mathbf{p}(\mathbf{p} \cdot \mathbf{E}_{PC}^r) + R_V[\mathbf{E}_{PC}^r - \mathbf{p}(\mathbf{p} \cdot \mathbf{E}_{PC}^r)] \tag{9.80}$$

with $\hat{\mathbf{r}}^r$ calculated as above, $\mathbf{p} = \mathbf{n} \times \mathbf{r}_i/|\mathbf{n} \times \mathbf{r}_i|$ and $\mathbf{E}_{PC}^r$ the reflected field for a perfectly conducting surface (calculated using Equation 9.79). The vertical and horizontal polarisation reflection coefficients (R_V and R_H) are given by

$$R_V = \frac{C\eta_r^{-2} - \sqrt{\eta_r^{-2} - S^2}}{C\eta_r^{-2} + \sqrt{\eta_r^{-2} - S^2}} \tag{9.81}$$

$$R_H = \frac{C - \sqrt{\eta_r^{-2} - S^2}}{C + \sqrt{\eta_r^{-2} - S^2}}, \tag{9.82}$$

where η_r ($= \eta/\eta_0$) is the relative impedance of the reflecting medium. Quantities C and S are given by

$$C = -\mathbf{n} \cdot \hat{\mathbf{r}}^i \quad \text{and} \quad S^2 = 1 - C^2. \tag{9.83}$$

Example An electromagnetic wave is incident upon the ground ($\epsilon_r = 10$ and $\sigma = 0.01$) at an angle ϕ from the normal. If the electric field component lies in the plane through the normal and the propagation direction ($\mathbf{E}_0^i \cdot \mathbf{p} = 0$), investigate the loss of power due to the reflection.

The loss (in decibels) after reflection will be given by

$$L = 10 \log_{10}(|\mathbf{E}_0^i|^2/|\mathbf{E}_0^r|^2), \tag{9.84}$$

where, since $\mathbf{E}_0^i \cdot \mathbf{p} = 0$,

$$|\mathbf{E}_0^r|^2 = |R_V|^2 |\mathbf{E}_{PC}^r|^2 = |R_V|^2 |\mathbf{E}_0^i|^2 \tag{9.85}$$

with reflection coefficient R_V given by

$$R_V = \frac{\cos\phi \, \eta_r^{-2} - \sqrt{\eta_r^{-2} - \sin^2\phi}}{\cos\phi \, \eta_r^{-2} + \sqrt{\eta_r^{-2} - \sin^2\phi}}, \tag{9.86}$$

where

$$\eta_r = \frac{\epsilon_r^{-\frac{1}{2}}}{\sqrt{1 - \frac{j\sigma}{\omega\epsilon}}} = \frac{\epsilon_r^{-\frac{1}{2}}}{\sqrt{1 - \frac{j}{0.0536f}}} \tag{9.87}$$

and f is the frequency in megahertz. At low frequencies (below a few megahertz) $\eta_r \ll 1$ and, except near grazing incidence, the ground will behave like a perfect conductor ($R_V \approx 1$). As grazing incidence is approached (ϕ towards $\pi/2$) the reflection coefficient R_V will change in sign and tend towards the value -1. The angle where this change takes place is known as the *Brewster angle*.

9.5   Guided wave propagation

Transmission lines are important in that they allow us to transfer electromagnetic energy between devices with very little loss. They do this by guiding electromagnetic waves along a confined propagation path. We have already discussed this process through a simple model based on lumped components and, for coaxial transmission lines (Figure 9.10), this gives results that are identical to those derived from the full electromagnetic theory. The full theory yields the fields

$$H_r = H_z = E_\theta = E_z = 0 \qquad (9.88)$$

$$E_r = -\frac{1}{r}\frac{V_0}{\ln(\frac{a}{b})}\exp(-j\beta z) \qquad (9.89)$$

$$H_\theta = \frac{I_0}{2\pi r}\exp(-j\beta z), \qquad (9.90)$$

where the voltage and current amplitudes are connected according to $V_0 = Z_0 I_0$ ($Z_0 = (\eta/2\pi)\ln(b/a)$ is the characteristic impedance of the line). Note that the electric field, magnetic field and propagation direction $\hat{z}$ are all mutually orthogonal. Furthermore, $E_r = \eta H_\theta$, where η is the impedance of the medium that separates the inner and outer conductors. Consequently, the wave travelling down the coaxial line has the character of a TEM wave. Strictly speaking, the microstrip variety of transmission line does not support a TEM wave. Its field structure is extremely complex but, for a substrate with

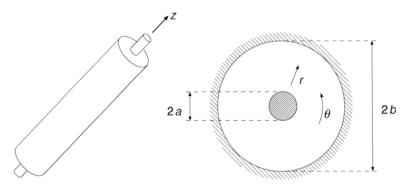

Figure 9.10   Coaxial waveguide.

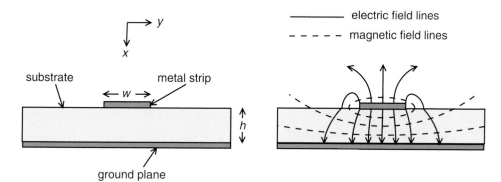

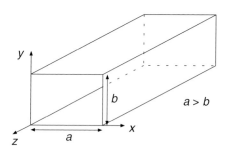

Figure 9.11   Microstrip geometry and fields.

Figure 9.12   Hollow rectangular waveguide.

thickness much less than a strip width, the field is approximately TEM with

$$E_x \approx \frac{V_0}{h} \exp(-j\beta z), \quad E_y \approx E_z \approx 0$$

$$H_y \approx \frac{I_0}{w} \exp(-j\beta z) \quad \text{and} \quad H_x \approx H_z \approx 0 \tag{9.91}$$

under the metal strip itself (note that $V_0 = (h\eta/w)I_0$ and $\beta = \omega/c$, where η and c are the impedance and propagation speed in the substrate). Outside this region, the field will adjust in a complex fashion to the exterior conditions (as illustrated in Figure 9.11).

There are other varieties of *waveguide* that do not even approximately support TEM waves. Consider the hollow rectangular waveguide that is shown in Figure 9.12 (we assume $a > b$). We need to find a propagating electromagnetic wave that has electric field with zero tangential components at the walls of the waveguide. The simple TEM wave

$$\mathbf{E} = E_0 \hat{\mathbf{y}} \exp(-j\beta z) \tag{9.92}$$

$$\mathbf{H} = H_0 \hat{\mathbf{x}} \exp(-j\beta z) \tag{9.93}$$

($H_0 = -E_0/\eta$) will only satisfy these conditions on the walls for which $y = 0$ and $y = b$ (not on the walls for which $x = 0$ and $x = a$). Consider, instead, an electromagnetic field that is composed of two plane waves travelling at angles $\pm\alpha$ to the z direction

$$\mathbf{E} = E_0\hat{\mathbf{y}}\exp[-j\beta(-x\sin\alpha + z\cos\alpha)] - E_0\hat{\mathbf{y}}\exp[-j\beta(x\sin\alpha + z\cos\alpha)]. \quad (9.94)$$

This field still satisfies the boundary conditions on sides $y = 0$ and $y = b$, but also exhibits a zero tangential electric field on side $x = 0$. It remains to ensure that the tangential electric field is zero on side $x = a$. This will be the case for an angle α such that $a\beta\sin\alpha = \pi$ and from which

$$\mathbf{E} = 2jE_0\hat{\mathbf{y}}\sin\left(\frac{\pi}{a}x\right)\exp(-j\beta_{10}z), \quad (9.95)$$

where $\beta_{10} = \beta\cos\alpha = \sqrt{\beta^2 - \pi^2/a^2}$. It will be noted that there is a *cut-off* frequency $\omega_c = \pi c/a$ below which the propagation constant β_{10} will become imaginary and not allow a propagating wave (the fields will be exponentially decaying in the propagation direction and are said to be *evanescent*). The phase speed of the composite wave is given by

$$V_p = \frac{\omega}{\beta_{10}} = \frac{c}{\sqrt{1 - \omega_{1m}^2}} \quad (9.96)$$

and from which it is seen to be dispersive. The magnetic field corresponding to the above electric field has the form

$$\mathbf{H} = -\frac{2\pi E_0}{\omega\mu a}\left[\frac{j\beta_{10}a}{\pi}\sin\left(\frac{\pi x}{a}\right)\hat{\mathbf{x}} + \cos\left(\frac{\pi x}{a}\right)\hat{\mathbf{z}}\right]\exp(-j\beta_{10}z) \quad (9.97)$$

($H_x = -E_y/Z_{10}$ where the waveguide impedance is calculated from $Z_{10} = \omega\mu/\beta_{10}$). It will be noted that the electric component is perpendicular to the propagation direction $\hat{\mathbf{z}}$, but the magnetic field is not. Such a wave is said to be transverse electric (TE). The above wave *mode* is only one of a series of TE modes (α satisfying $a\beta\sin\alpha = n\pi$ for any positive integer n will also produce electromagnetic fields that satisfy the requisite boundary conditions). In general, there is a double infinity of TE modes with cut-off frequencies given by

$$\omega_{lm} = c\pi\sqrt{\frac{l^2}{a^2} + \frac{m^2}{b^2}}, \quad (9.98)$$

where l and m are the positive integers that label the modes. The mode with cut-off frequency ω_{lm} is designated TE_{lm}, with Equation 9.95 expressing the electric field of a TE_{10} mode. It is clear that, with $a > b$, the TE_{10} mode has the lowest cut-off frequency of all modes. Besides the TE modes, there are also transverse magnetic (TM) modes for which the magnetic field is the only field perpendicular to the propagation direction. The cut-off frequencies of these modes are given by the same formula as for the TE modes, but both l and m must be greater than zero for a TM_{lm} mode to be non-trivial.

In general, the phase speed of a mode (both TE and TM) will satisfy

$$V_p = \frac{c}{\sqrt{1 - \omega_{lm}^2}} \qquad (9.99)$$

from which it can be seen that the problems of dispersion will be exacerbated if too many modes are present. Fortunately, it is possible to engineer a waveguide that only supports a TE_{10} mode and hence minimises the effects of dispersion. We need to choose dimensions such that ω_{10} is less than the required operating frequency ω and ω_{lm} is greater than the required operating frequency for all other modes (i.e., only the β_{10} propagation constant is totally real). It turns out that it is only necessary to ensure that $\omega > \omega_{10}$, $\omega < \omega_{20}$ and $\omega < \omega_{01}$ for this to be the case.

Example Design a rectangular waveguide that only supports a TE_{10} mode at a frequency of 2 GHz.

We will need to ensure that $\omega > \omega_{10}$, $\omega < \omega_{20}$ and $\omega < \omega_{01}$, where $\omega = 2\pi f = 2\pi \times 2 \times 10^9$ rad/s. From Expression 9.98 for the cut-off frequencies, we obtain $\omega_{10} = c\pi/a$, $\omega_{20} = 2c\pi/a$ and $\omega_{01} = c\pi/b$. Consequently $4 \times 10^9 > c/a$, $4 \times 10^9 < 2c/a$ and $4 \times 10^9 < c/b$. Since $c = 3 \times 10^8$ m/s, these inequalities will imply that $a > 7.5$ cm, $a < 15$ cm and $b < 7.5$ cm (note that the condition $a > b$ is automatically satisfied). If we choose $a = 10$ cm and $b = 5$ cm, all constraints will be satisfied and the resulting waveguide will only support a TE_{10} mode at 2 GHz. The question now arises as to whether there are other frequencies f for which the waveguide supports TE_{10} modes alone. Such frequencies must satisfy $f > c/2a$, $f < c/a$ and $f < c/2b$ and from which 1.5 GHz $< f < 3$ GHz.

Consider the electric field

$$\mathbf{E} = \frac{j}{2} E_0 \hat{\mathbf{y}} \sin\left(\frac{\pi}{a}x\right) [\exp(-j\beta_{10} z) - \exp(+j\beta_{10} z)] \qquad (9.100)$$

$$= E_0 \hat{\mathbf{y}} \sin\left(\frac{\pi}{a}x\right) \sin(\beta_{10} z) \qquad (9.101)$$

which is a combination of right and left travelling TE_{10} modes. It will be noted that the field is zero on the planes $z = 0$ and $z = \lambda_{10}/2$ ($\lambda_{10} = 2\pi/\beta_{10}$). As a consequence, we can replace these planes by metal plates without affecting the field. That is, a closed metal box of with dimensions $a \times b \times \lambda_{10}/2$ could contain an electric field of arbitrary magnitude at frequency ω_{10} (i.e., the box is resonant at frequency ω_{10}). Due to the multitude of modes that can exist inside a waveguide, it is obvious that the above mode (designated TE_{101}) is only one of a great number of resonant modes. It does, however, yield the lowest resonant frequency. A rectangular cavity can be turned into a one-port device through a small loop, or probe, placed inside the cavity (as shown in Figure 9.13). Around the cavity resonant frequency, the device will then behave as a series resonant LCR circuit with extremely high Q (several thousand).

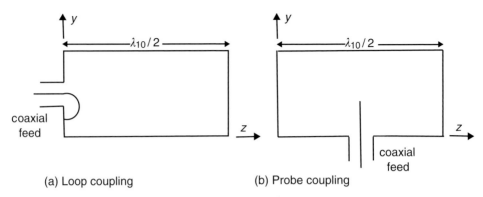

(a) Loop coupling   (b) Probe coupling

Figure 9.13   Rectangular cavities with loop and probe coupling.

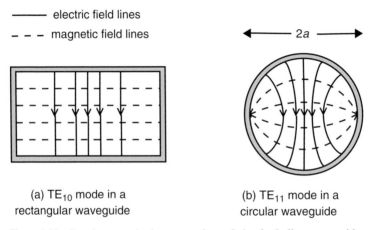

(a) TE_{10} mode in a
rectangular waveguide

(b) TE_{11} mode in a
circular waveguide

Figure 9.14   Dominant modes in rectangular and circular hollow waveguides.

As with rectangular hollow waveguides, circular hollow waveguides can also support an infinity of TE and TM modes. The mode with the lowest cut-off frequency is the TE mode that is designated TE_{11}. This has an angular cut-off frequency of $\omega_{11} = 1.841c/a$ with the next lowest cut-off frequency at $\omega_{21} = 3.054c/a$ (c is the electromagnetic propagation speed in the material that fills the waveguide and a is the waveguide radius). Figure 9.14 shows the electromagnetic field lines for a rectangular waveguide supporting a TE_{10} mode and a circular waveguide supporting a TE_{11} mode. It will be noted that the fields are very similar and, for this reason, such waveguides can be joined with very little leakage into other modes. The wavelength of a TE_{11} mode at frequency ω will be given by $\lambda_{11} = 2\pi c(\omega^2 - \omega_{11}^2)^{-1/2}$ and, as in the case of a rectangular guide, a $\lambda_{11}/2$ length can be closed off with metal ends to form a cavity with resonant frequency ω.

It is also possible to manufacture a waveguide without metal walls. Indeed, as we have already seen, an abrupt change in permittivity can cause a considerable amount of energy to be reflected. Consequently, a rod that has a large permittivity in comparison to

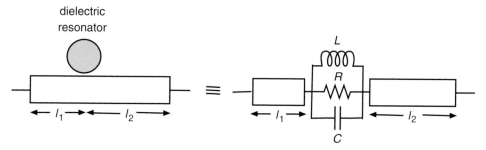

Figure 9.15   High Q bandstop microstrip formed with dielectric resonator.

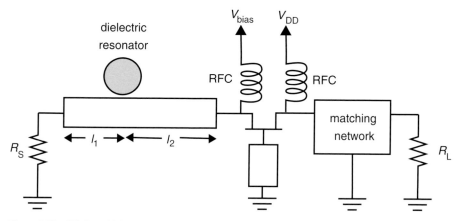

Figure 9.16   High stability oscillator that uses a dielectric resonator.

its surrounds will be extremely effective at trapping waves. The modes in such dielectric waveguides are quite complex and, in general, do not separate into TE and TM modes. (It is interesting to note that the dominant mode, that designated HE_{11}, has a cut-off frequency of zero.) As with the hollow variety, it is possible to fashion a resonator from a finite length of this waveguide and such a *dielectric resonator* (DR) can exhibit a Q of several thousand if suitable material is used. The dielectric waveguide modes have both internal and external fields and, although evanescent in nature, the external field can be used to couple energy. For example, a resonator can be placed in close proximity to a microstrip (as shown in Figure 9.15) and form a device that has a bandstop characteristic. This is equivalent to breaking the microstrip with an extremely high Q parallel-tuned LCR circuit. Such an arrangement is often used to produce extremely stable oscillators for gigahertz frequencies, as shown in Figure 9.16. At resonance, the length of transmission line to the right of the resonator will be open circuit and its length will be chosen to cancel the reactance of the transistor. The transistor will have negative input resistance due to the transmission line in its gate and will thus oscillate at the resonant frequency. Away from resonance, however, resistor R_S will stabilise the transistor and hence stop oscillations. Clearly, the quality of oscillations will depend upon the Q of the dielectric resonator.

9.6   Wave sources

An alternative means of studying magnetic fields is through the magnetic vector potential **A** for which

$$H = \frac{1}{\mu} \nabla \times A. \tag{9.102}$$

It is possible to define **A** such that it satisfies the driven *Helmholtz equation*

$$\nabla^2 A + \omega^2 \mu \epsilon A = -\mu J \tag{9.103}$$

and together with an electric field **E** of the form

$$E = -j\omega A + \frac{\nabla(\nabla \cdot A)}{j\omega\epsilon\mu} \tag{9.104}$$

this ensures that the time-harmonic Maxwell equations are satisfied. In the case of a plane wave,

$$A = A_0 \exp(-j\beta \, \hat{p} \cdot r), \tag{9.105}$$

where A_0 is a constant vector and $\hat{p}$ is a unit vector in the direction of propagation. The fields **E** and **H** will then take the form

$$E = -j\omega A + j\omega\hat{p}(A \cdot \hat{p}) \tag{9.106}$$

and

$$H = -\frac{1}{\eta}\hat{p} \times E. \tag{9.107}$$

The important thing to note is that field **A** satisfies an equation that has been extensively studied and for which there exists a large number of solution techniques. In particular, for a bounded source with current density **J**,

$$A(r) = \mu \int_V \frac{J(r')\exp(-j\beta \, |r - r'|)}{4\pi \, |r - r'|} \, dV', \tag{9.108}$$

where $dV' = dx' \, dy' \, dz'$ in Cartesian coordinates and V is a volume that contains the sources. Note that

$$|r - r'| = \sqrt{(r - r') \cdot (r - r')} = r\sqrt{1 - 2\frac{r \cdot r'}{r^2} + \frac{|r'|^2}{r^2}}, \tag{9.109}$$

where $r = \sqrt{r \cdot r}$. At large distances from a bounded source, this will imply the approximation $|r - r'| \approx r - r \cdot r'/r$. Consequently, to the leading order in $1/r$, the vector

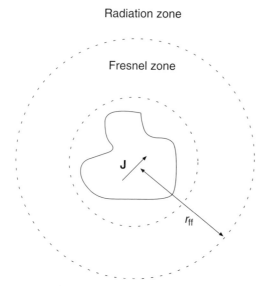

Radiation zone

Fresnel zone

J

r_{ff}

Figure 9.17   Field zones in relation to the radiating sources.

potential can be approximated by

$$\mathbf{A} = \frac{\mu}{4\pi r} \exp(-j\beta r) \int_V \mathbf{J}(\mathbf{r}') \exp(j\beta\,\hat{\mathbf{r}}\cdot\mathbf{r}')\,dV', \tag{9.110}$$

where $\hat{\mathbf{r}}$ is a unit vector in the radial direction. Locally, the field has the character of a plane wave moving in the radial direction ($\hat{\mathbf{p}} = \hat{\mathbf{r}}$) and so it is possible to use the plane wave expressions (9.106 and 9.107) to calculate the electric and magnetic fields. An important issue, however, is the region over which approximation 9.110 is valid. The region of validity is known as the *radiation* (or *Fraunhofer*) *zone* (see Figure 9.17). For sources with overall size D, the radiation zone is normally taken to be the region beyond a sphere of radius $r_{ff} = 2D^2/\lambda$ that is centred on the sources. This specific value of r_{ff} is that distance beyond which the phase error in approximation 9.110 is less than $\pi/8$. The region where the radiation dominates, but the distance is less than r_{ff}, is known as the *radiating near-field* (or *Fresnel*) *zone*. This zone starts at a distance of about $\sqrt{D^3/\lambda}$ from the source and the region inside this distance is known as the *reactive near-field zone*.

Example Electric charges $-q$ and q oscillate about the origin in the z-direction. Both oscillations have frequency ω and amplitude d ($d \ll \lambda$), but are 180° out of phase. The current vector will be given by

$$\mathcal{J} = -q\frac{d\mathbf{r}_-}{dt}\delta[\mathbf{r} - \mathbf{r}_-(t)] + q\frac{d\mathbf{r}_+}{dt}\delta[\mathbf{r} - \mathbf{r}_+(t)],$$

where $\mathbf{r}_+(t) = d\sin(\omega t)\hat{\mathbf{z}}$ and $\mathbf{r}_-(t) = -d\sin(\omega t)\hat{\mathbf{z}}$ are the position vectors of the charges (note that $\delta(\mathbf{r}) = \delta(x)\delta(y)\delta(z)$, where δ is the Dirac delta function). Find the approximate (assume $\delta(\mathbf{r} - \mathbf{r}_{\pm}(t)) \approx \delta(\mathbf{r})$) time harmonic current $\mathbf{J}$ ($\mathcal{J} = \Re\{e^{j\omega t}\mathbf{J}\}$) and hence find an expression for the electric field in the radiation zone.

Approximating the delta functions by $\delta(\mathbf{r})$, the full current density is given by

$$\mathcal{J} = 2qd\omega\cos(\omega t)\delta(\mathbf{r})\hat{\mathbf{z}} \tag{9.111}$$

and, since $\cos(\omega t) = \Re\{\exp(j\omega t)\}$, we will have

$$\mathbf{J} = 2qd\omega\delta(\mathbf{r})\hat{\mathbf{z}}. \tag{9.112}$$

Substituting the above current density into Expression 9.110, we obtain

$$\mathbf{A} \approx \frac{2\mu qd\omega}{4\pi r}\exp(-j\beta r)\hat{\mathbf{z}} \tag{9.113}$$

on noting that $\int_V \delta(\mathbf{r})A(\mathbf{r})\,dV = A(0)$. Then, from Expression 9.106 with $\hat{\mathbf{p}} = \hat{\mathbf{r}}$,

$$\mathbf{E} \approx -\frac{2j\mu qd\omega^2}{4\pi r}\exp(-j\beta r)(\hat{\mathbf{z}} - \hat{\mathbf{r}}\hat{\mathbf{z}}\cdot\hat{\mathbf{r}}) \tag{9.114}$$

which is valid for $r > 8d^2/\lambda$.

EXERCISES

(1) Decompose the plane electromagnetic wave

$$\mathbf{E} = \frac{E_0}{\sqrt{2}}(\hat{\mathbf{x}} + \hat{\mathbf{y}})\exp(-j\beta z)$$

into right- and left-hand circular polarised waves.

(2) Consider the electric field

$$\mathcal{E} = E_0[\cos(\omega_1 t - \beta_1 z) + \cos(\omega_2 t - \beta_2 z)]\hat{\mathbf{x}} \tag{9.115}$$

with components at frequencies at ω_1 and ω_2. If the medium has dispersion relation $\beta = (\omega/c)\sqrt{1 - \omega_c^2/\omega^2}$, where ω_c is a cut-off frequency, find an expression for the difference in time $\Delta\tau$ that it takes the two components to travel distance L. If ω_1 and ω_2 are the effective frequency limits of a pulse, how could you interpret $\Delta\tau$?

(3) A wall of thickness d ($\sigma = 0.01$ S/m, $\mu_r = 1$ and $\epsilon_r = 20$) is required to attenuate a 100 MHz signal by 20 dB; calculate the thickness of the wall.

(4) A 300 MHz radio signal must pass through a cavity wall (both cavity and wall sections are 10 cm thick). Calculate the attenuation of the wall if the solid parts have $\sigma = 0.1$ S/m, $\mu_r = 1$ and $\epsilon_r = 40$.

(5) Calculate the power flowing down a waveguide for the TE_{10} mode represented by Equation 9.95.

(6) The walls of a practical waveguide will be imperfectly conducting and so a TE_{10} mode will be attenuated as it propagates down a rectangular waveguide (cross-sectional dimensions a and b). Assuming that $a \ll b$, the main loss will occur on the walls of length b. Since a TE_{10} mode can be regarded as two plane waves that reflect back and forth between the waveguide walls, the loss can be calculated from the reflection losses. Use this approach to calculate the attenuation of such a mode at frequency ω after it has travelled a distance l. (Assume arbitrary values σ, μ and ϵ for the conductivity, permeability and permittivity of the waveguide walls.)

(7) An electric charge $-q$ orbits around charge q at angular speed ω in an orbit that is very much smaller than a wavelength. The current vector will be given by $\mathcal{J} = -q(d\mathbf{r}_0/dt)\delta(\mathbf{r} - \mathbf{r}_0(t))$, where $\mathbf{r}_0(t) = a\cos(\omega t)\hat{\mathbf{x}} + b\sin(\omega t)\hat{\mathbf{y}}$ is the position of the moving charge. Find the approximate (assume $\delta(\mathbf{r} - \mathbf{r}_0(t)) \approx \delta(\mathbf{r})$) time harmonic current $\mathbf{J}$ ($\mathcal{J} = \Re\{e^{j\omega t}\mathbf{J}\}$) and hence find an expression for the electric field in the radiation zone.

SOURCES

Edminster, J. A. 1993. *Electromagnetics* (2nd edn). Schaum's Outline Series. New York: McGraw-Hill.

Felsen, L. B. and Marcuvitz, N. 1994. *Radiation and Scattering of Waves*. IEEE Press and Oxford University Press.

Harrington, R. F. 1961. *Time-harmonic Electromagnetic Fields*. New York: McGraw-Hill.

Popovic, Z. and Popovic, B. D. 2000. *Introductory Electromagnetics*. Upper Saddle River, NJ: Prentice-Hall.

Pozar, D. M. 1998. *Microwave Engineering* (2nd edn). New York: John Wiley.

Smith, A. A. 1998. *Radio Frequency Principles and Applications*. Piscataway, NJ: IEEE Press.

10 Antennas

Antennas are the means by which electromagnetic wave energy is fed into, and extracted from, the propagation medium. They are a key element in RF systems and their design and analysis constitutes a very important area of RF engineering. The problems of antenna design are many and varied. Modern spread spectrum systems will require antennas that are capable of operating over a wide range of frequencies. Mobile communications have a requirement for small efficient antennas that radiate over a wide arc. Radar, on the other hand, requires antennas that illuminate only a narrow arc, but can be steered over a wide region. In many systems, the requirements turn out to be conflicting and it is important for the designer to understand the practical constraints in order to achieve the best compromise solution. The present chapter seeks to introduce the most fundamental concepts of antenna engineering and to describe some important varieties of antenna.

10.1 Dipole antennas

Broadly speaking, an antenna is a device that transforms wave propagation down a transmission line (a physically narrow channel) into wave propagation through free space (a physically wide channel) and vice versa. If we consider a parallel wire transmission line, we could conceive of opening it out at its end to better couple the waves into free space. The opened out section would then correspond to a dipole antenna. In its unopened state, the transmission line will reflect back to the source almost all of the energy that reaches its end. When opened out, however, the line allows the energy to spread out into space and there will only be partial reflections. (Figure 10.1 shows a snapshot of the electric field lines as they spread out into space.) The notion of a dipole as an opened out transmission line suggests, by analogy with transmission lines, that we can also picture the antenna in terms of a combination of inductors and capacitors (as shown in Figure 10.2). In this model, the capacitance steadily reduces as we move away from the dipole *feed* and the inductance steadily increases. The analogy with transmission lines also suggests that the current I will exhibit a sinusoidal distribution

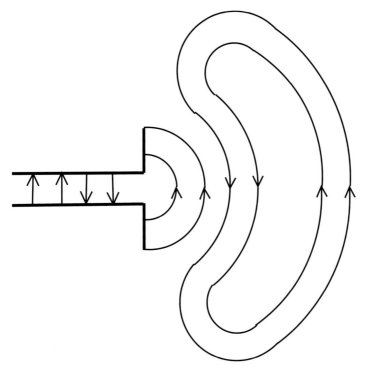

Figure 10.1   A snapshot of the electric field near a dipole.

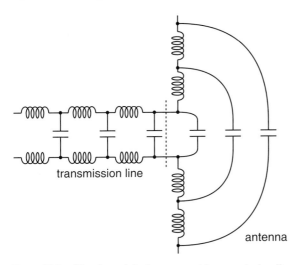

Figure 10.2   Circuit model of antenna with transmission line.

along the antenna wire with zero current at the open wire ends (see Figure 10.3 for some examples of current distributions).

It is convenient to introduce the notion of an *ideal dipole*, an antenna that consists of a short length h of uniform current I in a specified direction. The vector potential of

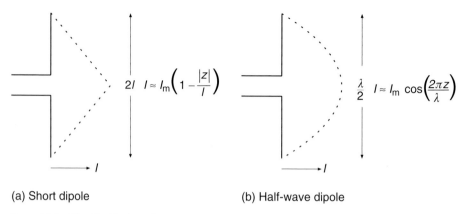

(a) Short dipole   (b) Half-wave dipole

Figure 10.3   The distribution of current on dipole antennas (λ is the free space wavelength).

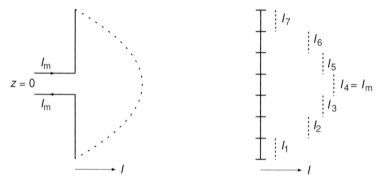

Figure 10.4   Half-wave dipole represented as a series of ideal dipoles.

such an element is given by

$$A = \frac{\mu I \mathbf{h}}{4\pi |\mathbf{r} - \mathbf{r}_i|} \exp(-j\beta |\mathbf{r} - \mathbf{r}_i|), \tag{10.1}$$

where $\mathbf{h}$, known as the *vector length*, is a vector of magnitude h in the direction of the current ($\mathbf{r}_i$ is the position of the dipole). Other more complex antennas can be regarded as a combination of such ideal elements. For example, the field of a half-wave dipole could be approximated by the sum of the fields due to N ideal dipole elements (Figure 10.4 shows the case where $N = 7$). Our major problem in analysing the behaviour of an antenna is that of obtaining an accurate estimate of the distribution of current (the sinusoidal distribution is only an approximation). This is difficult and, in general, requires the use of a computer model. Such models are normally based on an approximation such as that provided by a finite number of ideal dipoles. The currents on these dipoles are then adjusted such that the tangential electric field on the antenna surface is zero. Once the currents are known, the fields of the individual ideal dipoles can be calculated and summed to form the total field of the antenna.

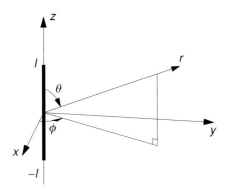

Figure 10.5 Dipole and coordinate system.

There are very few antennas that are amenable to exact analysis and numerical techniques, of the type described above, will normally be required. Some headway, however, can be made in the case of a dipole (Figure 10.5). For a z-directed dipole, we can assume the current distribution

$$I(z) = I_\mathrm{m} \frac{\sin(\beta(l - |z|))}{\sin \beta l}, \tag{10.2}$$

where I_m is the current at the centre ($z = 0$) and $2l$ is the dipole length. Far from the antenna, the vector potential will be given by

$$\mathbf{A} = \frac{\hat{\mathbf{z}}\mu}{4\pi r} \exp(-j\beta r) \int_{-l}^{l} I(z) \exp(j\beta z \, \cos\theta) \, dz \tag{10.3}$$

(note that the current distribution representing the antenna is given by $\mathbf{J} = I(z)\delta(x)\delta(y)\hat{\mathbf{z}}$ for $-l \leq z \leq l$ with $\mathbf{J} = 0$ otherwise). In the far field, the potential reduces to

$$\mathbf{A} \approx \frac{\mu I_\mathrm{m}}{2\pi \beta r} \exp(-j\beta r) f(\theta)\hat{\mathbf{z}}, \tag{10.4}$$

where

$$f(\theta) = \frac{\cos(\beta l \, \cos\theta) - \cos \beta l}{\sin^2 \theta \sin \beta l}. \tag{10.5}$$

Since the field will behave locally as a plane wave with propagation direction $\hat{\mathbf{r}}$ (the unit vector in the radial direction),

$$\mathbf{E} \approx -j\omega\mathbf{A} + j\omega(\mathbf{A} \cdot \hat{\mathbf{r}})\hat{\mathbf{r}} \tag{10.6}$$

$$\mathbf{H} \approx \frac{1}{\eta}\hat{\mathbf{r}} \times \mathbf{E}, \tag{10.7}$$

where η is the impedance of the propagation medium. In terms of polar coordinates, the above expressions yield

$$E_\theta \approx \frac{j\omega\mu I_\mathrm{m}}{2\pi \beta r} \exp(-j\beta r) f(\theta) \, \sin\theta \tag{10.8}$$

with $H_\phi = E_\theta/\eta$ and $H_\theta = H_\mathrm{r} = E_\mathrm{r} = E_\phi = 0$.

An important quantity is the average power P_{rad} that is radiated by the antenna. For a time-harmonic field, the Poynting vector can be averaged over a period to yield the expression $\frac{1}{2}\Re\{\mathbf{E} \times \mathbf{H}^*\}$ for the average power that flows through a unit area. The total power is obtained by integrating over a surface S that surrounds the antenna

$$P_{rad} = \frac{1}{2} \int_S \Re\{\mathbf{E} \times \mathbf{H}^*\} \cdot \hat{\mathbf{r}} \, dS. \tag{10.9}$$

Noting the relationship between $\mathbf{E}$ and $\mathbf{H}$, and taking S to be a sphere in the radiation zone, we obtain

$$P_{rad} = \int_0^{2\pi} \int_0^{\pi} \frac{\mathbf{E} \cdot \mathbf{E}^*}{2\eta} r^2 \sin \theta \, d\theta d\phi. \tag{10.10}$$

In the case of a dipole, the polar form of the electric field will yield

$$P_{rad} = \frac{1}{2\eta} \int_0^{2\pi} \int_0^{\pi} |E_\theta|^2 r^2 \sin \theta \, d\theta d\phi$$

$$= \frac{\eta |I_m|^2}{4\pi} \int_0^{\pi} f(\theta)^2 \sin^3 \theta \, d\theta. \tag{10.11}$$

For a half-wave dipole, $f(\theta) = \cos\left(\frac{\pi}{2} \cos \theta\right)/\sin^2 \theta$ and the above integral can be evaluated numerically to yield $P_{rad} = 1.218(\eta/4\pi)I_m^2$. In the case of a short dipole ($\beta l \ll 1$), $f(\theta) = \beta l/2$ and $P_{rad} = 40\pi^2(l/\lambda)^2 I_m^2$.

If all the power that is supplied to the antenna is radiated, then the antenna source will see a *radiation resistance* of value $R_{rad} = 2P_{rad}/I_m^2$. For a short dipole $R_{rad} = 80(\pi l/\lambda)^2$ (in ohms) and for a half-wave dipole $R_{rad} \approx 73\,\Omega$. The radiation resistance, however, is only part of the load that an antenna will present to its source. In addition, there will be a reactive part X_A which, for a half-wave dipole, is generally found to be around $43\,\Omega$ (i.e., a half-wave dipole is non-resonant). In reality, the antenna feed slightly disturbs the sinusoidal distribution of current and it is found that a dipole will achieve its first resonance ($X_A = 0$) at a length of around 0.47λ. A short dipole will be far off resonance and exhibit a capacitive reactance of $X_A = (-120/\beta l)(\ln(l/a) - 1)$ (in ohms) where a is the radius of the dipole rod.

The above considerations are ideal and any realistic antenna will dissipate energy as heat on its structure and in its surroundings. From previous considerations, the power dissipated in the antenna metal will be given by $P_{ohm} = (1/2\pi a)\int_{-l}^{l} (R_s I^2/2) \, dz$, where R_s is the surface resistance. Consequently, for a short dipole, $P_{ohm} = (l R_s/6\pi a)I_m^2$ from which the effective ohmic loss resistance will be $R_{ohm} = l R_s/3\pi a$. In total, the input impedance of an antenna will consist of the radiation resistance R_{rad}, the ohmic resistance R_{ohm} and the reactance X_A (i.e., $R_{rad} + R_{ohm} + jX_A$). The *effi-ciency* η_A of the antenna is a measure of the proportion of power that is lost as radiation

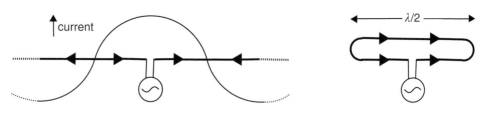

(a) Current distribution on an arbitrary length dipole   (b) Folded dipole

Figure 10.6   The folded dipole.

(good loss) as opposed to heating (bad loss). In the present context

$$\eta_A = \frac{P_{rad}}{P_{rad} + P_{ohm}} = \frac{R_{rad}}{R_{rad} + R_{ohm}}. \tag{10.12}$$

For a metal half-wave dipole that is located well away from dissipative structure, R_{ohm} is effectively zero and the efficiency approaches one (i.e., 100 per cent). For a short dipole, however, it will be noted that $R_{rad} \propto l^2$ and $R_{ohm} \propto l$ and this means that the efficiency approaches zero as l approaches zero. A short dipole can therefore be extremely inefficient, a property it holds in common with many antennas that are small when compared to a wavelength. An additional downside to a small antenna is its large Q (a result due to Wheeler states that $Q \geq \alpha^{-3}\beta^{-3}$, where α is the radius of the smallest sphere that contains the antenna). This indicates a severe limitation on bandwidth, a major problem in this age of spread spectrum communications. The Q of an antenna can be calculated according to $Q = |X_A/(R_{rad} + R_{ohm})|$ and, for a short dipole, this leads to a result that is consistent with that of Wheeler.

For dipoles that are resonant, an interesting variant is known as the *folded dipole*. In general, the current distribution on a resonant dipole will be sinusoidal (as shown in Figure 10.6a). If each arm of a one wavelength dipole is folded back at a quarter wavelength from the feed, these folded arms meet at a point where the current amplitude has a maximum. Furthermore, the folding process reverses the direction of the current and hence causes a distribution on the folded section that is identical to that on the half wavelength containing the feed. As a consequence, the radiated power is $P_{rad} = 1.218\,(\eta/\pi)I_m^2$ and this, in turn, implies a radiation resistance of $R_{rad} = 292\,\Omega$. Although the gain of the folded dipole is identical to that of the standard half-wave dipole, this variety of dipole does have the advantage of a much increased bandwidth.

The dipoles discussed so far are known as *electric dipoles* since, close to the antenna, their behaviour is dominated by the electric field component. There are, however, antennas for which the fields close to the antenna are predominantly magnetic in nature. The simplest of these is known as a *magnetic dipole* and has a vector potential of the form

$$\mathbf{A} = j\beta\mu\,\frac{SI\hat{\mathbf{r}} \times \mathbf{n}}{4\pi r}\,\exp(-j\beta r). \tag{10.13}$$

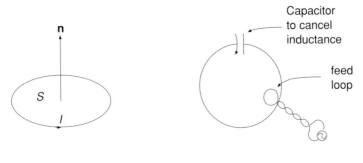

(a) Loop parameters   (b) Realistic loop antenna

Figure 10.7   Magnetic dipole antenna and practical realisation.

Such a dipole can be realised as a small (dimensions $\ll \lambda$) loop that carries a current I and encloses a plane area S with unit normal $\mathbf{n}$ (see Figure 10.7a). The loop will have a radiation resistance of $R_{rad} = 20(\beta^2 S)^2$, an ohmic resistance of $R_{ohm} = (l/2\pi a)R_s$ and an inductive reactance of $X_A = (\omega\mu_0 l/2\pi)(\ln(4l/\pi a) - 1.75)$, where l is the loop circumference and a is the radius of the loop wire. As with other small antennas, ohmic losses are a major problem and their construction needs to ensure the lowest possible conductor resistance (high conductivity metal and large cross-section conductor). The antenna will normally include a capacitor in order to cancel out the loop inductance and can be fed through a small inductive probe (see Figure 10.7b).

Example   A small circular loop (1 m in diameter) is required to operate at 30 MHz (the loop is made of aluminium tube with $\sigma = 4 \times 10^7$ S/m). Find the radius of tube that ensures a 150 kHz bandwidth. Calculate the antenna efficiency and find the value of capacitance that would be needed to resonate the loop.

The circumference of the loop will be $l = \pi$ metres and the frequency $\omega = 6\pi \times 10^7$ rad/s. Consequently, the loop will have reactance $X_A = 120(\ln(4/a) - 1.75)$. Since $\beta = \omega/c = 0.63$, we will have a radiation resistance of $R_{rad} = 20\beta^4 S^2 = 1.94\,\Omega$. Furthermore, since the surface resistance will have the value $R_S = \sqrt{\omega\mu_0/2\sigma} = 0.0017\,\Omega$ at 30 MHz, the ohmic resistance of the loop will be given by $R_{ohm} = lR_S/2\pi a = 0.000\,85/a$. A 150 kHz bandwidth will require a Q of 30 MHz/0.15 MHz $= 200$ and, since $Q = |X_A|/(R_{rad} + R_{ohm})$, we will need $|X_A| \le 200(R_{rad} + R_{ohm})$ or

$$120\left[\ln\left(\frac{4}{a}\right) - \frac{7}{4}\right] \le 200\left(1.944 + \frac{0.000\,85}{a}\right). \tag{10.14}$$

This inequality will be satisfied for radii above $a = 3$ cm and for which $R_{ohm} = 0.028\,\Omega$. Consequently,

$$\text{efficiency} = \frac{R_{rad}}{R_{rad} + R_{ohm}} = \frac{1.944}{1.944 + 0.028} = 0.986. \tag{10.15}$$

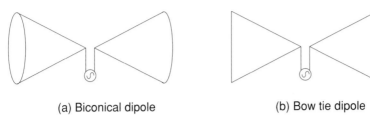

(a) Biconical dipole   (b) Bow tie dipole

Figure 10.8   Biconical and bow tie dipoles.

Now the reactance of the loop will have the value $X_A = 120(\ln(4/0.03) - 1.75) =$ $377 \, \Omega$ and this will require a capacitive reactance of $X_C = -377 \, \Omega$ for resonance. Since $X_C = -1/\omega C$, this will mean a capacitance of $C = 1/(2\pi \times 3 \times 10^7 \times 377) = 14.1 \, \text{pF}$.

Bandwidth is often a critical issue with antennas due to the fact that many RF systems will need to operate over a large range of frequencies. As we have already seen, the input impedance can vary markedly with frequency and it can be difficult to devise a suitably flexible matching circuit. Ideally, we would like an antenna that exhibits a constant impedance over the required frequency range. A possible solution comes from the *Rumsey principle*, which states:

If, by arbitrary scaling, an antenna is transformed into a structure equal to the original one, its properties will be independent of frequency.

Examples of such structures are the biconical and bow tie dipole antennas shown in Figure 10.8. Unfortunately, the Rumsey result requires a structure of infinite extent and the dipoles will need to be truncated for practical implementation. If the length of dipole is a half wavelength at the lowest operating frequency, the resulting structure will be more or less frequency independent at frequencies above this. Such dipoles do not necessarily need to be solid structures and can be constructed out of a limited number of metal rods that spread out from the feed. The input impedance of a biconical antenna will be $120 \, \ln(\cot(\theta_h/2))$ (in ohms), where θ_h is the angle that the cone subtends to its axis (i.e., the input impedance can be manipulated by the cone angle).

10.2   Effective length and gain

In the radiation zone, all antennas will exhibit the same radial dependence $\exp(-j\beta r)/r$ and this can be factored out. Clearly, the radiation behaviour of an antenna is characterised by a function of direction alone. An important angular characterisation is afforded by a vector quantity known as the *effective length*. In the radiation zone, the electric field of an ideal dipole has the form

$$\mathbf{E} = \frac{j\omega\mu I_m}{4\pi} \mathbf{h}_{\text{eff}} \frac{\exp(-j\beta r)}{r}, \qquad (10.16)$$

where the effective length is defined by $h_{eff} = (\hat{r} \cdot h)\hat{r} - h$ and h is the vector length of the dipole. The radiation field of a general antenna can be expressed in the above form, but h will now be a complex function of direction

$$h = \frac{1}{I_m} \int_V J(r') \exp(j\beta\hat{r} \cdot r') \, dV',$$
(10.17)

where V is a volume that contains the antenna and I_m is the current in the antenna feed. When the antenna is represented by a distribution of ideal dipoles, the effective length can be calculated according to

$$h_{eff} = \sum_i h^i_{eff} \exp(j\beta\hat{r} \cdot r_i),$$
(10.18)

where h^i_{eff} is the effective length of the ith dipole and r_i is its position. As we will see later, the effective length also provides information about the properties of an antenna in receive mode. In particular, when immersed in electromagnetic field E, an antenna will exhibit an open-circuit voltage of the form

$$V_{OC} = h_{eff} \cdot E,$$
(10.19)

where h_{eff} is the value of the effective length in the direction of the field source. From this expression, we can see the need for a receiving antenna to be matched to the polarisation of an incoming electric wave.

Although the effective length is an extremely useful description of antenna behaviour, antenna radiation performance is more commonly described in terms of its directivity (or the related concept of gain). Directivity is defined in terms of radiation from an isotropic antenna, an antenna that radiates equally well in all directions. An isotropic antenna is impossible to achieve in practice, but represents a useful standard against which to measure the performance of a real antenna. If an isotropic antenna radiates a total power P_{rad}, it will radiate $P_{rad}/4\pi$ into a unit solid angle. For a real antenna that radiates total power P_{rad}, we define its directivity $D(\theta, \phi)$ in a particular direction to be the power radiated into a unit solid angle in that direction when scaled upon $P_{rad}/4\pi$, that is

$$D(\theta, \phi) = \frac{r^2 E(r, \theta, \phi) \cdot E(r, \theta, \phi)^* / 2\eta}{P_{rad}/4\pi}$$
(10.20)

in the limit $r \to \infty$. In the case of a z-directed half-wave dipole,

$$D(\theta, \phi) = \frac{5}{3} \left[\frac{\cos\left(\frac{\pi}{2} \cos\theta\right)}{\sin\theta} \right]^2$$
(10.21)

and, in the case of a short dipole,

$$D(\theta, \phi) = \frac{3}{2} \sin^2\theta.$$
(10.22)

A small loop antenna with the z-axis normal to the plane of the loop will have directivity

$$D(\theta, \phi) = \frac{3}{2} \sin^2 \theta \qquad (10.23)$$

which is the same result as for an electric dipole!

In reality, we are more interested in the relation between the total power P_a accepted by an antenna from its source and the power radiated in a particular direction. Consequently, we define the *gain* function $G(\theta, \phi)$ by

$$G(\theta, \phi) = \frac{r^2 \mathbf{E}(r, \theta, \phi) \cdot \mathbf{E}(r, \theta, \phi)^*/2\eta}{P_a/4\pi} \qquad (10.24)$$

in the limit $r \to \infty$. This is a characterisation that incorporates the ohmic losses that occur on the antenna and in its surrounds. As a consequence, gain represents a more realistic measure of antenna performance. Since the difference between P_a and P_{rad} is caused by ohmic losses, G and D are related through $G(\theta, \phi) = \eta_A D(\theta, \phi)$.

The gain is often represented geometrically by its *gain pattern*. This is a three-dimensional surface that surrounds the origin (associated with the antenna). In a particular direction, the distance of the surface from the origin is the value of the gain in that direction. (The distance is often expressed in a decibel scale in order to allow for a wide range of gain values.) The gain surface could possibly be represented as a shaded plot, as shown in Figure 10.9 for the case of a dipole, but it is difficult to derive accurate gain values from such a representation. An alternative is to display the gain by means of representative cross-sectional cuts through the gain surface. Figures 10.10, 10.11 and 10.12 show such gain cross-sections for short, half wavelength and 5/4 wavelength dipoles, respectively. Since the patterns are symmetric about the

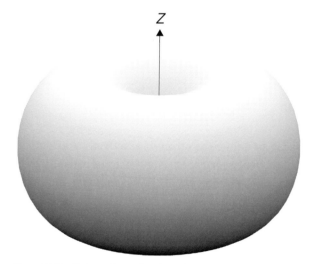

Figure 10.9   Gain surface of a short dipole.

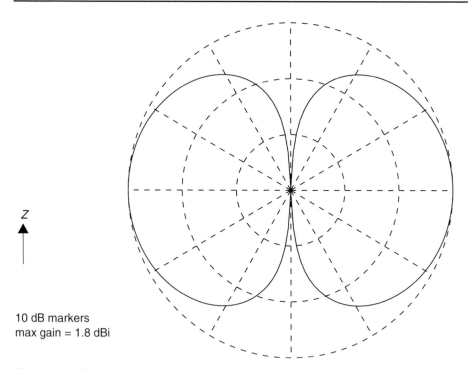

10 dB markers
max gain = 1.8 dBi

Figure 10.10　Gain of a short dipole ($f = 150\,\text{MHz}$) in a cross-section through the dipole axis.

dipole axis, a representative cut has been taken through this axis. It will be noted that the gain patterns have been scaled upon the maximum value of gain (indicated on the plot). The patterns of the short and half-wave dipoles are very similar, but the 5/4 wavelength dipole has an increased maximum gain and exhibits four *side lobes* in addition to the two *main lobes*. It should be noted that gain is sometimes quoted as a single number, usually in terms of decibels over the isotropic case (dBi). In such circumstances, the single number is taken to refer to the maximum gain of the antenna.

10.3　The monopole antenna

In electrostatics it is possible to mimic the effect of a perfectly conducting plane by replacing it with a system of image charge. It is possible to extend this concept to radiating current sources by replacing the conducting plane with a system of image currents. Figure 10.13 shows three current elements and their equivalent image system (note the change in direction for horizontal components of image current).

A monopole antenna is a vertical conducting rod that is placed above a conducting plane and fed against this plane. By the image concept, the plane can be replaced

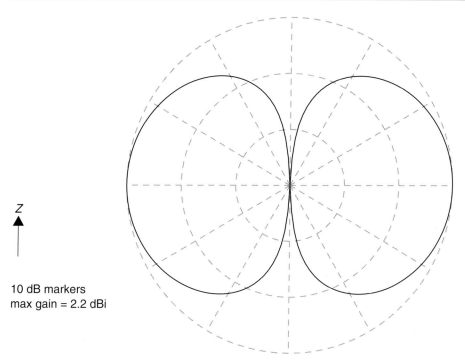

z

10 dB markers
max gain = 2.2 dBi

Figure 10.11   Gain pattern of a half wavelength dipole ($f = 150\,\text{MHz}$).

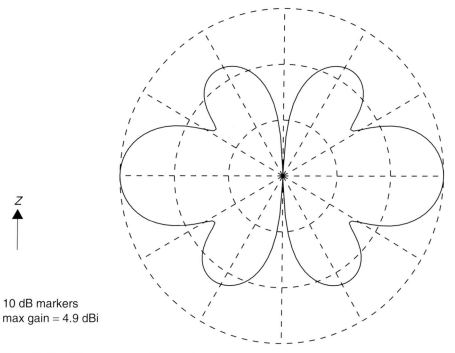

z

10 dB markers
max gain = 4.9 dBi

Figure 10.12   Gain pattern of a 5/4 wavelength dipole ($f = 150\,\text{MHz}$).

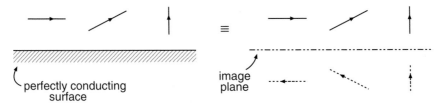

Figure 10.13 Image current concept.

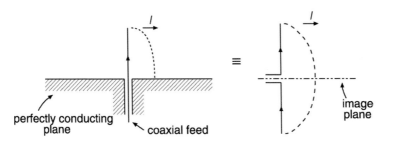

Figure 10.14 Connection between monopole and dipole antennas.

by an image monopole and the equivalent system will therefore consist of a dipole (Figure 10.14). As a consequence, the properties of a monopole antenna can be directly inferred from those of a dipole antenna. Firstly, the image plane cuts the antenna feed in half and so the feed voltage for a monopole will be half that of the related dipole, but with the same maximum current. As a consequence, the input impedance of a monopole $Z_{\mathrm{in}}^{\mathrm{mono}}$ will be half that of the corresponding dipole impedance $Z_{\mathrm{in}}^{\mathrm{dip}}$,

$$Z_{\mathrm{in}}^{\mathrm{mono}} = \frac{V_{\mathrm{in}}^{\mathrm{mono}}}{I_{\mathrm{in}}^{\mathrm{mono}}} = \frac{\frac{1}{2} V_{\mathrm{in}}^{\mathrm{dip}}}{I_{\mathrm{in}}^{\mathrm{dip}}} = \frac{1}{2} Z_{\mathrm{in}}^{\mathrm{dip}}. \tag{10.25}$$

In the case of a quarter-wave monopole,

$$Z_{\mathrm{in}} = 36.5 + 21.5\mathrm{j} \tag{10.26}$$

and, for resonance, the monopole will need to be about 0.235 wavelengths long. The directivity of a monopole will also change from that of a dipole. For directions above the conducting plane, the radiated power will be the same as for a dipole, but the total radiated power will be half. Consequently, for a quarter-wave monopole, the maximum directivity will be 5.2 dBi (3 dB more than that of a dipole).

It is often impossible to deploy a monopole that is a full quarter wave in length (quite unacceptable in the case of a mobile phone) and so it is important to find ways of shortening the antenna. We could simply use a short monopole, together with a matching network, in order to produce the required antenna input impedance. As we have already seen, however, this can produce an extremely inefficient antenna with very narrow bandwidth. For an antenna consisting of a conducting rod, the physical length for resonance is determined by the wave speed v along the rod (a physical wavelength

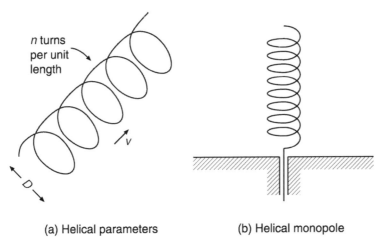

(a) Helical parameters   (b) Helical monopole

Figure 10.15   Shortening a monopole antenna.

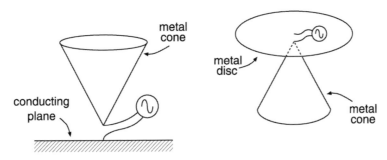

Figure 10.16   Broadband conical and discone antennas.

is given by $2\pi v/\omega$). For an uncoated rod, v is the speed of light in the propagation medium. It is possible, however, to lower the wave speed on the antenna by using a helical winding instead of a straight rod. For a helical winding with a diameter D that is much smaller than a wavelength (see Figure 10.15), the wave speed v can be crudely estimated from $2c/(2 + n\pi D)$, where n is the number of turns per unit length. Therefore, by utilising a helical coil instead of a rod, it is possible to obtain a very short antenna that is resonant.

In situations where a broadband antenna is required, the biconical and bow tie dipole designs can be split at the feed to provide broadband monopoles. In some circumstances, the ground plane of a conical monopole is truncated to a metal disc (radius about that of the cone base) in order to produce what is known as the discone antenna (Figure 10.16). Once again, the conducting surfaces of such an antennas can be realised as a series of rods that spread out from the feed.

Thus far, we have assumed that a monopole antenna is driven against a perfectly conducting plane. This, however, is an ideal that can never be realised. In reality, a

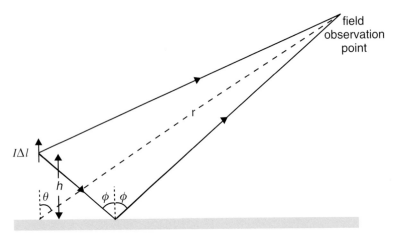

Figure 10.17   Ground reflections from a dipole antenna element.

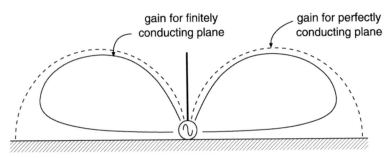

Figure 10.18   The effect of a finitely conducting plane.

conducting plane will have finite conductivity and so will absorb some of the monopole radiation (effectively reducing the magnitude of the image current). It is possible to consider the effect of the plane in terms of the reflection of waves generated by the current element above the plane. For radiation from a small current element of length Δl (see Figure 10.17), the field will take the form

$$E_\theta = \frac{j\omega\mu I \,\Delta l}{4\pi r} \exp(-j\beta r)[\exp(j\beta h \cos\theta) + R_V \exp(-j\beta h \cos\theta)]\sin\theta, \quad (10.27)$$

where R_V is the plane wave reflection coefficient (see Chapter 9) at the appropriate angle of incidence ϕ ($\phi \approx \theta$ for an observation point well above the current element) and h is the height of the element above the plane. (Note that the phase factors involving h are required to take account of the different distances travelled by the direct and reflected radiation.) The effect of the plane is dramatic at low elevations where the field, maximum in the case of a perfectly conducting plane, is now reduced to very low levels (see Figure 10.18).

The major problem for a real monopole antenna is the large amount of power that can be absorbed by the surface against which it is fed (i.e., it can have very low efficiency).

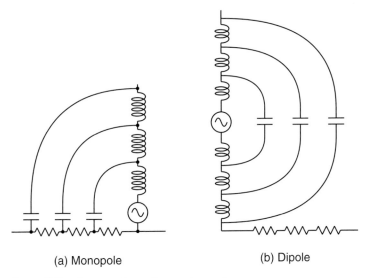

(a) Monopole   (b) Dipole

Figure 10.19   Antennas located over a lossy plane.

In circuit terms, this is modelled as shown in Figure 10.19a. Therefore, for the capacitive components of the antenna, the return path is through the resistors that represent the surface loss. This loss can be substantially reduced by placing a conducting mat (or radial wires) beneath the antenna. If space permits, however, it is much better to use a full dipole since the capacitive return paths will not have to traverse a lossy medium (see Figure 10.19b).

10.4   Feeding an antenna

One of the major problems in RF engineering is the efficient connection of the various elements that constitute an RF system (a transmitter and antenna, for example). For maximum power transfer, this will require a conjugate match between the system elements. Such matching could present a significant problem in a complex system and, for this reason, components tend to be designed around standard input and output impedances (usually 50 Ω). System components will normally contain matching circuits that ensure they present the standard impedance at their ports. In this fashion, the components can be connected by a standard transmission line (often 50 Ω coaxial cable).

Example   A transmitting system operates at 30 MHz and is required to use a 1 m (2 cm diameter) aluminium ($\sigma = 4 \times 10^7$ S/m) rod for its antenna. Design an L-network that will match the antenna to the 50 Ω impedance of the transmitter and its connecting transmission line.

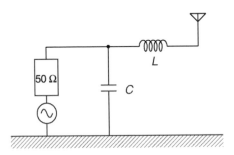

Figure 10.20 Matching a short monopole to a standard load.

If we consider the monopole to be fed against a perfectly conducting ground plane, the input impedance Z_A of the monopole will be half that of the equivalent dipole. The rod is short compared to a quarter wavelength and so the input impedance will be

$$Z_A = R_{rad} + R_{ohm} + jX_A = 40 \left(\frac{\pi l}{\lambda}\right)^2 + \frac{l R_s}{6\pi a} - j\frac{60}{\beta l}\left(\ln\left(\frac{l}{a}\right) - 1\right), \quad (10.28)$$

where $l = 1\,m$, $\lambda = 10\,m$, $a = 0.01\,m$, $\beta = \pi/5m^{-1}$ and $R_s = \sqrt{\frac{\omega\mu}{2\sigma}} = 0.0017\,\Omega$ for aluminium at 30 MHz. It is clear that the ohmic resistance is negligible and so

$$Z_A = R_{rad} + jX_A = 3.95\,\Omega - j344.27\,\Omega. \quad (10.29)$$

Consider the matching network shown in Figure 10.20. To match the 50 Ω source to this antenna, we first need to choose C such that the real part of 50 $\Omega \parallel 1/j\omega C$ is equal to the antenna resistance of 3.95 Ω (note that 50 $\Omega \parallel 1/j\omega C = (0.02 - j\omega C)/(0.0004 + \omega^2 C^2)$). At a frequency of 30 MHz, C will need have a value of 362 pF. Together, the source plus capacitor will exhibit a reactance of $-13.5\,\Omega$ (from the imaginary part of 50 $\Omega \parallel 1/j\omega C$) which must be cancelled by the series combination of the antenna reactance ($-344.27\,\Omega$) and the inductor reactance (ωL). We will need $-13.5 - 344.27 + \omega L = 0$ from which $L = 1.9\,\mu H$. It will be noted that the calculations thus far have assumed the ground to be a perfectly conducting plane. In the case of a realistic ground, however, the capacitive reactance of the antenna will be much larger (two or three times the size). This arises due to the large resistance in the return path through the ground (see the circuit model of Figure 10.19) which reduces the effective capacitance of the antenna.

When connecting by means of a standard transmission line, other properties besides impedance need to be considered. Care needs to be taken to ensure that the line can handle the power that will inevitably be dissipated in its structure. Furthermore, the question arises as to whether the line should be balanced or unbalanced. A parallel wire transmission line is geometrically symmetric and will therefore produce a balanced electric field between its conductors. On the other hand, a coaxial cable (and a microstrip line) will produce unbalanced fields between the conductors. Crudely speaking, a dipole

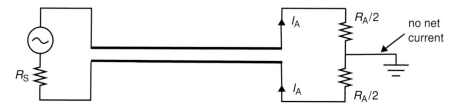

(a) Balanced antenna with twin wire feed

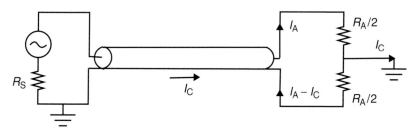

(b) Balanced antenna with coaxial feed

Figure 10.21   A balanced antenna when fed by (a) balanced and (b) unbalanced transmission lines.

antenna can be regarded as a parallel wire transmission line that has been opened out and so this line is the natural dipole feed. On the other hand, a monopole antenna can be regarded as a coaxial cable that has been opened out (the cable outer becoming the ground plane) and so this cable is its natural feed. The problem comes when we want to feed a balanced antenna (such as a dipole) with an unbalanced feed (such as a coaxial cable). Since coaxial cable is a fairly standard item, this situation is quite common. We might expect to be able to connect the cable inner conductor to one arm of a dipole and the outer to the other arm. The trouble with this configuration is that the outer conductor presents two possible paths for the current. The inner surface provides a path that is associated with the normal feeder function, but the outer surface provides an alternative path to the ground. This will result in an imbalance between the currents on the arms and, even worse, the current that flows down the cable outer can radiate in an undesired fashion (see Figure 10.21). The solution is to choke off the current flow down the cable outer, or to use a BALUN (balanced to unbalanced) transformer of the sort described in Chapter 7. Figure 10.22a shows a dipole with a quarter-wave transmission line choke (the outer of the cable is now part of a quarter-wave shorted stub) that will present an infinite impedance at the dipole feed. The same principle can be used to achieve a ground independent vertical antenna, as shown in Figure 10.22b. This antenna is known as a choke dipole and is extremely useful because of its compactness. The lower half of the dipole, together with the outer of the feed, forms a quarter-wave choke that prevents the flow of current down the cable outer. All current return must occur through the inter-arm capacitance and so normal dipole operation prevails. The lower dipole

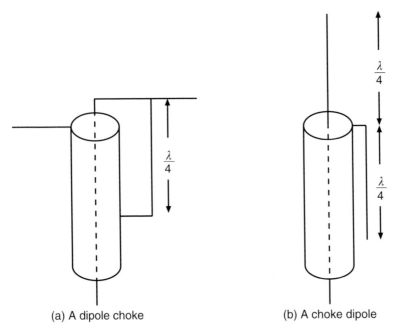

(a) A dipole choke   (b) A choke dipole

Figure 10.22   Dipoles with coaxial feeds.

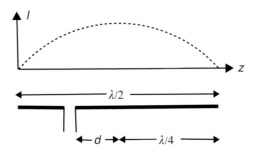

Figure 10.23   A half-wave dipole with offset feed.

arm (the one that forms the choke) will often consist of a suitably wide metal sleeve that is joined to the coaxial cable outer at the dipole centre.

Up to now we have always assumed our dipoles to have their feed point at the centre, but this is not mandatory. It is possible to place the feed at any point along the dipole, but there will be an associated change in antenna input impedance (Figure 10.23). The centre corresponds to the lowest possible impedance since this is the point at which current is greatest and voltage least. In general, the input impedance is given by

$$R_{in} \approx \frac{R_{centre}}{\cos^2 \beta d},$$ (10.30)

where d is the distance from the centre of the dipole. This expression would suggest that the impedance becomes infinite at the dipole ends. In reality, however, this value will be finite (although still very large).

At the end of the day, it is almost impossible to obtain an exact match between antenna and feed. Consequently, it is important to take the mismatch efficiency

$$\eta_M = 1 - |\Gamma|^2 \qquad (10.31)$$

into account in system planning. Γ (always less than 1 for a passive device) is the reflection coefficient at the junction between the feed and the antenna. It should be noted that the combined effect of mismatch, antenna loss and cable loss can often be quite considerable.

10.5 Array antennas

In the current crowded RF environment, it is often necessary to limit the directions in which energy is transmitted and to reduce the gain in the direction of interfering signals in the case of reception. Increasingly, there is a requirement for *smart* antennas that can change their gain pattern in real time. It is possible to increase gain, and to manipulate the shape of the gain pattern, by combining several antennas with suitable drive level and phase relationship. To see how this is achieved, we first look at the combination of two antennas in receive mode (reciprocity ensures that the same conclusions concerning gain will also hold in the transmit case).

Consider a pair of parallel dipoles (see Figure 10.24) separated by a distance l and connected to a combiner by different length cables (s_b and s_f). For a signal from the front (electric field E_f aligned to the dipoles), the combiner output voltage is given by

$$V_f = E_f h_{eff} \exp(-j\beta s_f) + E_f h_{eff} \exp[-j\beta(l + s_b)], \qquad (10.32)$$

where h_{eff} is the effective length of the dipole. (Note that s_b and s_f are assumed to be in terms of electrical length, the length in free space that would cause the same phase change.) In the case of a signal from the back (electric field E_b aligned to the dipoles), the combiner output voltage is given by

$$V_b = E_b h_{eff} \exp(-j\beta s_b) + E_b h_{eff} \exp[-j\beta(l + s_f)] \qquad (10.33)$$

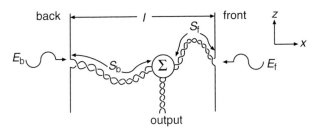

Figure 10.24   A pair of phased dipoles.

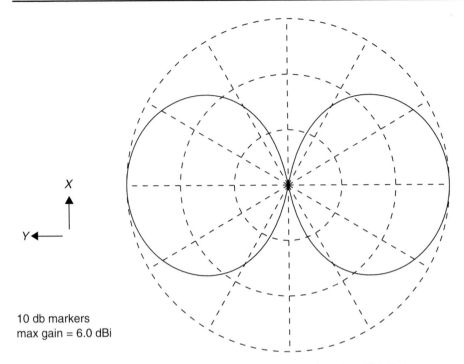

X

Y

10 db markers
max gain = 6.0 dBi

Figure 10.25   Gain pattern of two dipoles in broadside configuration ($f = 150$ MHz).

and for signals from the side (electric field E_s aligned to the dipoles) by

$$V_s = E_s h_{eff} \exp(-j\beta s_b) + E_s h_{eff} \exp(-j\beta s_f). \tag{10.34}$$

Consider the case where $s_f = s_b$ (in phase contributions from front and back), then

$$V_f = [1 + \exp(-j\beta l)]E_f h_{eff} \exp(-j\beta s_b) \tag{10.35}$$
$$V_b = [1 + \exp(-j\beta l)]E_b h_{eff} \exp(-j\beta s_b) \tag{10.36}$$

and

$$V_s = 2E_s h_{eff} \exp(-j\beta s_b). \tag{10.37}$$

There is maximum response to signals from the sides and equal response to signals from front and back. In fact, for a spacing of $l = \lambda/2$, there is zero response from the front and back. The simulated gain pattern for two half-wave dipoles (spacing $l = \lambda/2$ and in phase feeds) is shown in Figure 10.25 (cross-section through a plane perpendicular to the dipole axes) and confirms the above simple theory.

Now consider the case where $s_f = s_b + (\lambda/2)$ (anti-phase contributions from front and back), then

$$V_f = [\exp(-j\beta l) - 1]E_f h_{eff} \exp(-j\beta s_b) \tag{10.38}$$
$$V_b = [1 - \exp(-j\beta l)]E_b h_{eff} \exp(-j\beta s_b) \tag{10.39}$$

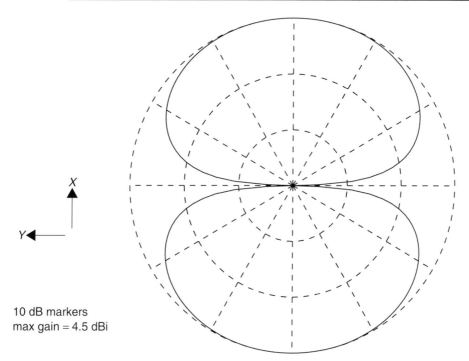

Figure 10.26   Gain pattern of two dipoles in endfire configuration ($f = 150$ MHz).

and

$$V_s = 0. \tag{10.40}$$

Note that there is now zero response from the side. Furthermore, when the spacing is $l = \lambda/2$ there will be maximum response to signals from front and back. The simulated gain pattern for two half-wave dipoles (spacing $l = \lambda/2$ and anti-phase feeds) is shown in Figure 10.26 and confirms the above simple theory.

In both of the above examples it will be noted that, by choosing the element spacing to be half a wavelength, we have achieved maximum response along one axis and zero response in an orthogonal plane. For spacings greater than a half a wavelength, however, the gain patterns have a tendency to break up with strong lobes in many directions. For element spacings less than half a wavelength, however, a much greater degree of directionality can be achieved. For quarter wavelength spacing and a 90° phase difference in the feeds ($s_f = s_b + \lambda/4$), the simple theory predicts zero response from the back and strong response in other directions. The simulated gain pattern for two half-wave dipoles in this configuration is shown in Figure 10.27. It should be noted, however, that the close spacing of the dipoles has caused mutual interactions that have degraded the response that is predicted by the simple theory (the response to the rear is reduced, but certainly not to zero). Element interaction is a serious source of degradation in array antennas and needs to be taken into account in their design. Figure 10.28 shows the mutual impedance between parallel and collinear half-wave dipoles

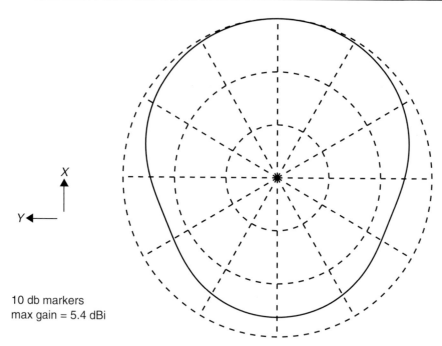

X

Y

10 db markers
max gain = 5.4 dBi

Figure 10.27 Gain pattern of dipoles with 90° phasing and λ/4 spacing ($f = 150$ MHz).

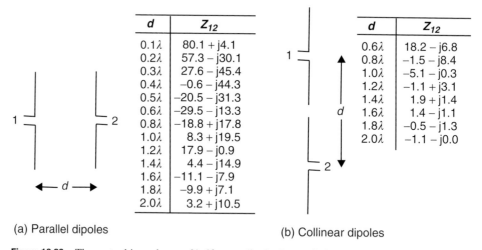

d	Z_{12}
0.1λ	80.1 + j4.1
0.2λ	57.3 − j30.1
0.3λ	27.6 − j45.4
0.4λ	−0.6 − j44.3
0.5λ	−20.5 − j31.3
0.6λ	−29.5 − j13.3
0.8λ	−18.8 + j17.8
1.0λ	8.3 + j19.5
1.2λ	17.9 − j0.9
1.4λ	4.4 − j14.9
1.6λ	−11.1 − j7.9
1.8λ	−9.9 + j7.1
2.0λ	3.2 + j10.5

d	Z_{12}
0.6λ	18.2 − j6.8
0.8λ	−1.5 − j8.4
1.0λ	−5.1 − j0.3
1.2λ	−1.1 + j3.1
1.4λ	1.9 + j1.4
1.6λ	1.4 − j1.1
1.8λ	−0.5 − j1.3
2.0λ	−1.1 − j0.0

(a) Parallel dipoles

(b) Collinear dipoles

Figure 10.28 The mutual impedance of half-wave dipoles in parallel and collinear configurations.

for a variety of spacings. It will be noted that, for spacings that are less than a quarter wavelength, the parallel elements have mutual impedance that is almost at the level of the self-impedance. The interaction is far less for the collinear dipoles, but is still quite substantial.

In changing the relative phasing of the feeds from 0° to 180°, we have *steered* the maximum antenna response from *broadside* (orthogonal to the line of antennas) to

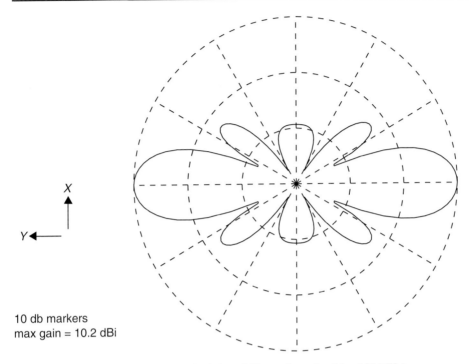

X

Y

10 db markers
max gain = 10.2 dBi

Figure 10.29 Gain pattern of five dipoles in broadside configuration ($f = 150$ MHz).

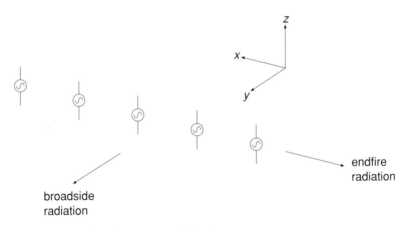

broadside
radiation

endfire
radiation

Figure 10.30 A five-element array of dipoles.

endfire (in the direction of the line of antennas). The ability to form a strong response in a particular direction (*beam forming*) and then to steer this response (*beam steering*) is an extremely important capability for modern radar and communication systems. Figure 10.29 shows the gain pattern for a line of five linear in-phase dipoles (see Figure 10.30), spaced $\lambda/2$ apart. The gain is plotted for a plane perpendicular to the dipole axes (the horizontal plane). As expected from the two dipole case, the in-phase feeds have produced a broadside response. It will be noted that the *beam width* (angular

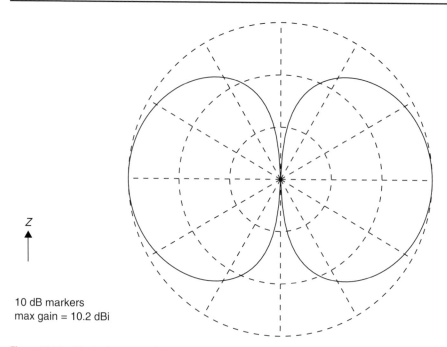

Z

10 dB markers
max gain = 10.2 dBi

Figure 10.31   Vertical pattern of broadside configuration ($f = 150$ MHz).

width over which the main radiation lobe is no more than 3 dB below its peak) has decreased and the peak gain has increased. In general, an increase in the length of an array will results in a decrease in beam width and an increase in gain. (For a large broadside array of N dipoles with spacing d, the beamwidth is approximately λ/Nd radians and the maximum directivity approximately $2Nd/\lambda$ times that for a dipole.) Figure 10.31 shows the gain in the vertical plane through broadside and from which it will be noted that the array has retained the vertical gain pattern of the dipole. Figure 10.32 shows the gain pattern of the same linear array, but with a phase increment of 180° between consecutive dipoles. It will be noted that the 180° phase increments result in an endfire pattern. It is possible to further manipulate the gain pattern with a decrease in element spacing and phase increment. Figure 10.33 shows the situation in which the spacing between the array elements has been reduced to a quarter wavelength and there is a phase decrement of 90° between elements in the x-direction. There is now an endfire pattern with strong gain in the x-direction (the direction can be reversed by replacing the 90° phase decrement by a 90° increment).

In general, the radiation field of an array of N identical antennas at positions $\mathbf{r}_1$ to $\mathbf{r}_N$ is given by

$$\mathbf{E} \approx \sum_{n=1}^{N} \frac{j\omega\mu I_m^n}{4\pi} \mathbf{h}_{\text{eff}} \frac{\exp(-j\beta|\mathbf{r} - \mathbf{r}_n|)}{|\mathbf{r} - \mathbf{r}_n|}, \tag{10.41}$$

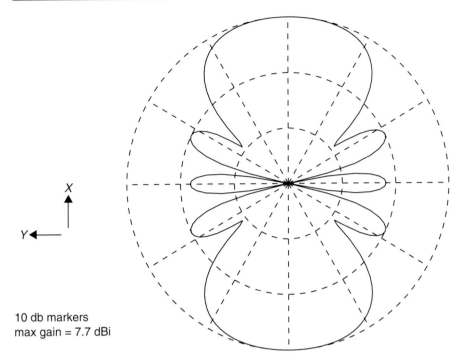

10 db markers
max gain = 7.7 dBi

Figure 10.32   Gain pattern of five dipoles in endfire configuration ($f = 150$ MHz).

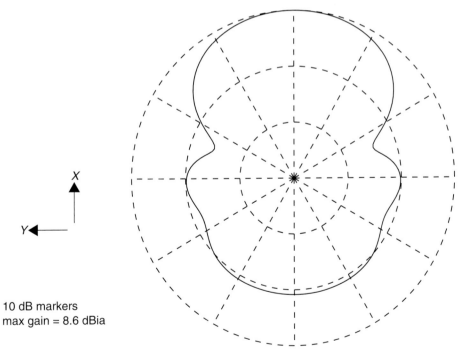

10 dB markers
max gain = 8.6 dBia

Figure 10.33   Gain pattern of five endfire dipoles with reduced spacing and phase increment ($f = 150$ MHz).

where the I_m^n is the feed current in the nth antenna and $\mathbf{h}_{\text{eff}}$ is its effective length. From the approximation $|\mathbf{r} - \mathbf{r}_n| \approx r - \mathbf{r} \cdot \mathbf{r}_n / r$, we obtain

$$\mathbf{E} \approx \frac{j\omega\mu}{4\pi} \mathbf{h}_{\text{eff}} \frac{\exp(-j\beta r)}{r} \sum_{n=1}^{N} I_m^n \exp(j\beta \, \hat{\mathbf{r}} \cdot \mathbf{r}_n) \tag{10.42}$$

at large distances from the array. Now consider the special case of antennas that are located along the x-axis with equal spacing d. If these antennas are driven by currents that have equal magnitude and a phase increment of α between adjacent antennas ($I_m^n = I_m^0 \exp\{j[n - (N+1)/2]\alpha\}$ and $\mathbf{r}_i = d[n - (N+1)/2)\hat{\mathbf{x}}]$, the electric field will reduce to

$$\mathbf{E} \approx \frac{j\omega\mu I_m^0}{4\pi} \mathbf{h}_{\text{eff}} \frac{\exp(-j\beta r)}{r} \sum_{n=1}^{N} \exp\left[j \left(n - \frac{N+1}{2} \right) (d\beta \, \cos \psi + \alpha) \right]$$

$$= \frac{j\omega\mu I_m^0}{4\pi} \mathbf{h}_{\text{eff}} \frac{\exp(-j\beta r)}{r} f(d\beta \, \cos \psi + \alpha), \tag{10.43}$$

where $f(\chi) = [\sin(N\chi/2)]/[\sin(\chi/2)]$ and ψ is the angle between the x-axis and the direction of the observation point (note that $\cos \psi = \sin \theta \, \cos \phi$ in terms of polar coordinates). This shows that the electric field of the array is simply the electric field of a single element multiplied by an *array factor* f that is a function of $\chi = d\beta \, \cos \psi + \alpha$ (this is the *pattern multiplication principle*). At $\chi = 0$ the array factor $f(\chi)$ will have a distinct peak (height N) and the width of this peak will narrow as N increases (see Figure 10.34). In other words, the radiation will be concentrated in a beam that narrows as the number of array elements increases. Furthermore, it is clear that the direction of this beam can be steered by varying the phase increment α between elements.

Figure 10.35 shows an example of an endfire array of dipoles that is fed from a single source. Moving out from the feed there is an effective phase increment of around 90° between elements which results in an endfire array with strong gain in the direction of the antenna source. It will be noted that the element lengths increase in size as we move away from the source. As a consequence, at a particular frequency, only those dipoles that are almost resonant will be active and this will limit the antenna gain to of the order of 10 dBi. On the plus side, however, the antenna will be able to provide a resonant structure over a wide range of frequencies. The lower frequency limit will be set by the resonant frequency of the longest dipole and the highest frequency by the resonant frequency of the shortest dipole. This antenna is known as a log periodic dipole array (LPDA) and is an example of the *extended Rumsey principle*:

If a structure, upon scaling by some ratio τ, becomes equal to itself it will have the same properties at frequencies f and τf.

A log periodic antenna is formed by choosing a value of τ and building a structure that reproduces itself on each scaling. Unfortunately, the Rumsey result only applies to an infinite structure and a real antenna will need to be truncated in order to achieve a

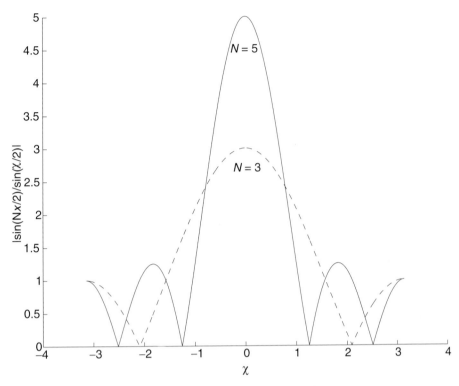

Figure 10.34   Array factors for linear arrays with equally spaced elements.

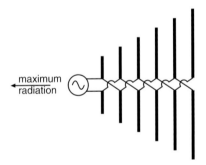

Figure 10.35   Log periodic dipole antenna.

practical design (such as in the case of the LPDA above). Providing τ is not too different from 1 (values of around 0.9 are normally favoured) the antenna will effectively be frequency independent from its lowest to its highest resonances, the limits being dictated by the size of the largest and smallest elements, respectively. The input impedance of the LPDA can be adjusted by variations in the characteristic impedance of the transmission line that joins the dipoles. Figure 10.36 shows a typical LPDA gain pattern in a plane perpendicular to the antenna and Figure 10.37 shows the corresponding pattern in the plane of the antenna.

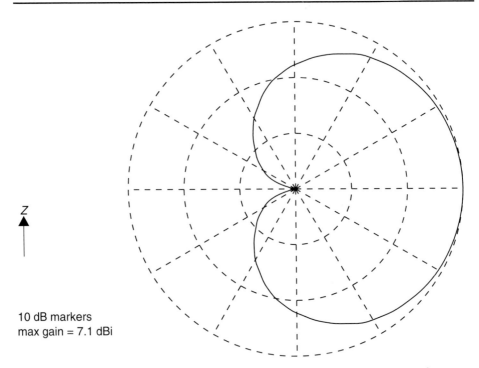

10 dB markers
max gain = 7.1 dBi

Figure 10.36 Typical gain pattern of an LPDA in a plane perpendicular to the antenna ($f = 30$ MHz).

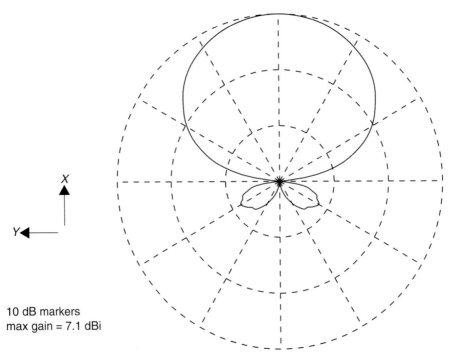

10 dB markers
max gain = 7.1 dBi

Figure 10.37 Typical gain pattern in the plane of the LPDA ($f = 30$ MHz).

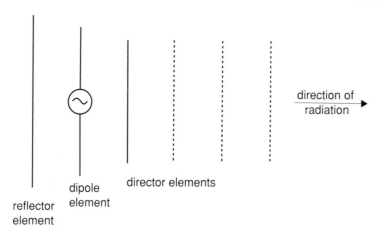

Figure 10.38 A general Yagi–Uda parasitic array antenna.

Another rather interesting single feed array is known, after its inventors, as the Yagi–Uda antenna. This antenna consists of a single driven dipole that has a *reflector* element to the rear and *director* elements to the fore (see Figure 10.38). The driven dipole induces currents in the reflector and director elements and these have phases such that the total configuration acts as an endfire array of dipoles. Consider the interaction between the driven dipole and one of the *parasitic* elements. In terms of the impedance matrix

$$V_1 = Z_{11}I_1 + Z_{12}I_2$$
$$0 = Z_{21}I_1 + Z_{22}I_2, \tag{10.44}$$

where subscript 1 refers to the driven element (driven by a source that imposes a voltage V_1 across the antenna terminals) and subscript 2 refers to a parasitic element regarded as short-circuited dipole (I_1 and I_2 refer to the respective currents in these elements). From the above equations, the current in a parasitic element will be given by $I_2 = -(Z_{21}/Z_{22})I_1$ ($Z_{22} \approx 73\,\Omega$). Noting the values of mutual impedance in Figure 10.28, it can be seen that an element spacing of around 20 per cent of a wavelength will cause parasitic element currents of similar magnitude to those in the driven element. There is obviously 180° of phase difference between the elements plus an additional amount due to the complex part of the mutual impedance. The complex part of Z_{22}, and hence the element phasing, can be adjusted by varying the lengths of the parasitic elements. An endfire pattern will normally require the reflector element to be slightly longer than the driven element and the director elements to be slightly shorter. Element lengths and spacing will need to be adjusted for optimal performance, but gains of over 8 dB can be obtained with just three elements (considerably higher gains with additional director elements).

10.6   Travelling wave antennas

For most antennas, the currents flowing into the feed point will travel out as waves across the antenna until reflected back at its extremities. This is a similar situation to that of an open-ended transmission line. As with the transmission line, the phase of the reflections will depend upon the distance of the extremities. Only when the antenna is excited at a specific frequency will the ingoing and outgoing currents combine to give a real input impedance. This resonant impedance will consist of the resistances caused by radiation and ohmic loss. As the frequency moves away from resonance, the currents will become increasingly out of phase and the input impedance will become increasingly reactive. In the case of a transmission line, however, the reflected current can be removed by loading the extremity with a resistor equal to the characteristic impedance Z_0 of the line. As a consequence, the line will present an input impedance Z_0 at all frequencies. Since an antenna can be viewed as a transmission line that has been opened out, it is reasonable to expect that a similar approach could produce an antenna with frequency independent input impedance. Figure 10.39 shows an antenna, known as a Beverage, that is based on this notion. The Beverage antenna consists of a wire (height h above the ground) that, together with the ground plane, forms a transmission line. This line is loaded with its characteristic impedance ($R \approx 138 \log_{10}(4h/d)$, where d is the wire diameter) at the opposite end to the antenna feed. Figure 10.39 also shows a typical vertical gain pattern for such an antenna. (The off-vertical pattern is found the by rotating the vertical pattern about the horizontal axis and the main lobe is at an angle $\alpha \approx \cos^{-1}(1 - 0.372\lambda/L)$ to this axis.) Since the current waves will only travel in one direction on the antenna, the Beverage is an example of what is often known as a *travelling wave* antenna. The wire of the antenna need not necessarily be parallel to the ground for the system to work. Indeed, a large variety of travelling wave antennas can be made by varying the wire geometry (by raising the height at its centre or even replacing the ground image by a real wire). The essential ingredient, however, is that the antenna has some mechanism for preventing reflected currents.

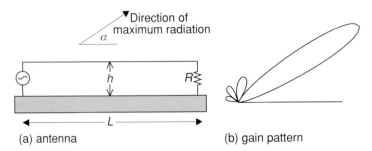

Figure 10.39   Beverage travelling wave antenna.

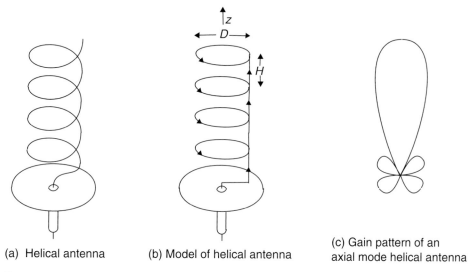

(a) Helical antenna   (b) Model of helical antenna   (c) Gain pattern of an
                                                                axial mode helical antenna

Figure 10.40   Helical antenna construction and gain.

Another antenna, essentially of the travelling wave variety, is a helical antenna operating in the axial mode. We have already seen that a helical antenna with a small diameter will slow wave progress along its structure and hence make a resonant monopole
(or dipole) of shorter length than the equivalent rod structure. In this case, the major
radiation is in a direction orthogonal to the axis and the helix is said to operate in the
normal mode. If the length of a single helix turn is of the order of a wavelength, however, the direction of maximum radiation will be along the helix axis and the antenna
is said to operate in the *axial mode*. Figure 10.40a shows a typical axial mode helix.
It will be noted that the antenna has a small ground plane and an unbalanced coaxial
feed. Each turn of the helix radiates power and so, by the end of the helix, there will be
very little power in the forward direction and hence very little reflection. Figure 10.40b
shows a simplified model of the antenna. It consists of an array of simple elements,
each consisting of a loop and a short monopole. For helix turns of the order of a wavelength, there will be a phase increment of about 180° between elements and hence,
from previous considerations, an endfire pattern of the form shown in Figure 10.40c
will result.

It is instructive to consider the electric field for the situation in which the dimensions
of the loop, and dipole, are relatively small when compared to a wavelength. In this
case

$$\mathbf{E} = \frac{j\omega\mu I_{\mathrm{m}} H}{4\pi} \frac{\exp(-j\beta r)}{r} \sin\theta\,\hat{\boldsymbol{\theta}} + \frac{\eta\beta^2 I_{\mathrm{m}} D^2}{16} \frac{\exp(-j\beta r)}{r} \sin\theta\,\hat{\boldsymbol{\phi}}, \quad (10.45)$$

where $\hat{\theta}$ and $\hat{\phi}$ are the unit vectors in the angular coordinate directions. It will be noted
that the orthogonal components of the field are 90° out of phase and so the radiation

of this antenna will not be linearly polarised. In fact, if $D = \sqrt{2H\lambda/\pi^2}$, the radiation from the helix will be circular polarised and this makes it an extremely useful device for combating the polarisation rotation that can be a feature of satellite signals. It should be noted, however, that the sense of the polarisation will be dictated by the sense of the helical winding.

10.7   Aperture antennas

Thus far we have concentrated on metallic wire structures as radiating elements, but the open end of a waveguide can also act as an antenna. Consider a section of waveguide (dimensioned such that only a TE_{10} mode propagates) with one end closed and the other open (see Figure 10.41). We can excite this waveguide by means of a quarter-wave monopole that is placed a distance $\lambda_{10}/4$ from the closed end ($\lambda_{10} = 2\pi/\beta_{10}$ is the wavelength based on the propagation constant β_{10} of a TE_{10} mode). The position of the monopole will ensure that the reflected TE_{10} mode is in phase with the direct TE_{10} mode in order to promote constructive interference (there is a $180°$ phase change at reflection together with $180°$ that is accrued during the propagation). At the open end of the waveguide, the emerging TE_{10} mode will spread out to form a radiation field and this can be analysed by means of an equivalent current distribution. We replace the waveguide by a conducting plate (the same size as the aperture) and illuminate it with a TE_{10} mode travelling in the opposite direction to the one in the waveguide. The reflected field will have the same structure as that emerging from the end of the waveguide. On the plate there will be a surface current density $\mathbf{J}_s = \mathbf{n} \times \mathbf{H}$, where $\mathbf{H}$ is the magnetic field at the plate surface and $\mathbf{n}$ is the unit normal on the plate (see Figure 10.41). The incident and reflected electric fields will cancel at the plate ($\mathbf{n} \times \mathbf{E} = 0$ on a perfect conductor) and this will imply a total magnetic field $\mathbf{H}$ that is double the incident magnetic field $\mathbf{H}_a$. As a consequence, the current on the plate will be given by $\mathbf{J}_s = 2\mathbf{n} \times \mathbf{H}_a$ and this will

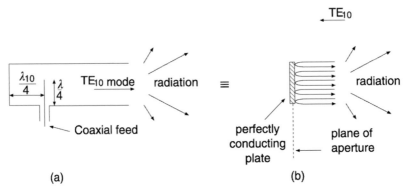

Figure 10.41   Waveguide antenna and analysis.

radiate a field that is equivalent to that generated by the aperture. At large distances, this will result in the vector potential

$$\mathbf{A} = \frac{\mu \exp(-j\beta r)}{4\pi r} \int_A \mathbf{J}_s(r') \exp(j\beta \hat{\mathbf{r}} \cdot \mathbf{r}') \, dS', \tag{10.46}$$

where A is the area of the aperture. The electric field will therefore be given by

$$\mathbf{E} = \frac{j\omega\mu \exp(-j\beta r)}{2\pi r} \int_A \{\hat{\mathbf{r}}\hat{\mathbf{r}} \cdot [\mathbf{n} \times \mathbf{H}_a(\mathbf{r}')] - \mathbf{n} \times \mathbf{H}_a(\mathbf{r}')\} \exp(j\beta \hat{\mathbf{r}} \cdot \mathbf{r}') \, dS' \tag{10.47}$$

and this will be strongest when $\hat{\mathbf{r}}$ is orthogonal to the aperture ($\hat{\mathbf{r}} = \mathbf{n}$). As a consequence, from Equation 10.20, the maximum directivity is given by

$$D_{\max} = \frac{\eta}{2} \frac{|\frac{\beta^2}{4\pi^2}| |\int_A \mathbf{n} \times \mathbf{H}_a(\mathbf{r}) \, dS|^2}{P_{\mathrm{rad}}/4\pi}, \tag{10.48}$$

where P_{rad} is the total radiated power. For the incident field, we will assume that the electric component $\mathbf{E}_a$ is approximately related to the magnetic component through the plane-wave expression $\mathbf{E}_a = \eta \mathbf{n} \times \mathbf{H}_a$. Since the power emerging from the waveguide will be the same as P_{rad}, we will therefore have that $P_{\mathrm{rad}} = (1/2\eta) \int_A |\mathbf{E}_a(\mathbf{r})|^2 \, dS$. The maximum directivity can now be expressed as

$$D_{\max} \approx \frac{4\pi}{\lambda^2} \frac{|\int_A \mathbf{E}_a(\mathbf{r}) \, dS|^2}{\int_A |\mathbf{E}_a(\mathbf{r})|^2 \, dS}, \tag{10.49}$$

where $\mathbf{E}_a$ is now to be interpreted as the electric field in the aperture. (It should be noted that Equation 10.49 is a general expression for the maximum directivity of an aperture antenna.) For a TE_{10} mode in a rectangular waveguide (sides of length a and b) we will have $\mathbf{E}_a = E_0 \hat{\mathbf{y}} \sin[(\frac{\pi}{a})x] \exp(-j\beta_{10} z)$ and, as a consequence,

$$D_{\max} = \frac{4\pi}{\lambda^2} \frac{|\int_0^b \int_0^a \sin(\frac{\pi}{a}x) \, dx \, dy|^2}{\int_0^b \int_0^a \sin^2(\frac{\pi}{a}x) \, dx \, dy}$$

$$= \frac{32ab}{\pi \lambda^2}. \tag{10.50}$$

Clearly, the directivity is proportional to the area of the aperture and the gain of a waveguide antenna is often increased by opening out its end to form a *horn* antenna (see Figure 10.42). A further increase in aperture area, and hence in directivity, can be achieved by placing the horn at the focus of a metallic parabolic reflector.

10.8 Patch antennas

We have already seen that a resonant wire can form an effective radiating structure. Cavities can also resonate when suitably excited and will radiate if provided with a suitable aperture. A simple cavity radiator can be formed from two parallel plates with

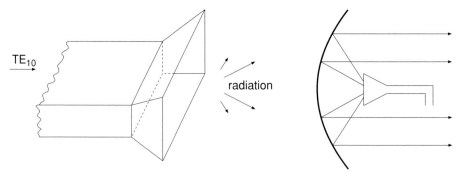

Figure 10.42 A horn antenna and a reflector antenna.

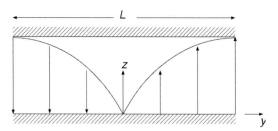

Figure 10.43 Electric field between plates at resonance.

only a small separation. The configuration will then radiate from its edges. Consider a rectangular plate with largest dimension L (Figure 10.43), then there will be a resonant mode at frequency

$$f_r = \frac{c_0}{2L\sqrt{\epsilon_r}}, \qquad (10.51)$$

where c_0 is the speed of light in a vacuum and ϵ_r is the relative permittivity of any material between the plates. The electromagnetic field between the plates can be approximated by

$$E_z \approx E_0 \sin\left(\frac{\pi y}{L}\right), \quad E_x \approx E_y \approx 0$$

$$H_x \approx \frac{j}{\omega\mu}\frac{\pi}{L}E_0\cos\left(\frac{\pi y}{L}\right) \quad \text{and} \quad H_y \approx H_z \approx 0. \qquad (10.52)$$

It will be noted that the tangential components of electric field are zero on the plates and that the tangential components of magnetic field are zero on the apertures. This last condition is known as the *magnetic wall condition* and is an approximate boundary condition that is appropriate to open circuits. From the formulas for the radiation fields of an aperture, it will be noted that the plate ends corresponding to $y = \pm L/2$ will be the only ones contributing to the radiation. These radiating ends will behave as a pair of y-directed electric dipoles a distance L apart. We will have, therefore, a two-element

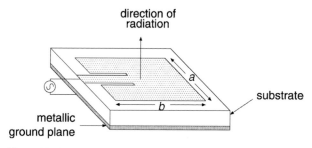

direction of
radiation

a

b

substrate

metallic
ground plane

Figure 10.44   A practical patch antenna with microstrip feed.

broadside array with maximum radiation along the z-axis. A practical implementation of the above antenna is shown in Figure 10.44 and is known as a *patch* antenna. It consists of a rectangular conducting patch that is separated from a ground plane by a substrate of relative permittivity ϵ_r. To ensure that the mode represented by Equation 10.52 is dominant, we will need a patch width a that is less than the patch length b ($\approx c_0/2f\sqrt{\epsilon_r}$) and substantially greater than the substrate thickness h. The design is fed by means of a microstrip line that terminates at a point in the patch which exhibits the impedance of the line. At a distance d from the edge, a patch will present an impedance R_{in} given by

$$R_{in} = R_{edge} \cos^2\left(\frac{\pi}{b}d\right), \tag{10.53}$$

where R_{edge} is the impedance at the edge of the patch (this can be approximated by $60\lambda_0/a$, where λ_0 is the free space wavelength). The fact that the impedance drops to zero at the centre of the patch introduces another option. If we remove the half of the antenna without a feed, and replace this by a short circuit (from the patch to the ground plane), we obtain an antenna that is half the size and still resonant.

Patch antennas are easily implemented using printed circuit techniques and for this reason are sometimes known as *printed* antennas. If the ground screen extends for a significant region beyond the edges of the patch, the dipoles representing the radiation properties will be reflected in this screen. As a consequence, there will be enhanced radiation in the positive z-direction and it is possible for a half-wave patch to achieve over 8 dBi of gain.

Example   Design a patch antenna for operation at 1 GHz with input impedance 50 Ω. The antenna should use a 3 mm FR4 substrate ($\epsilon_r = 4.4$).

The frequency $f_r = 1\,\text{GHz}$ will require a length $b = c_0/2f\sqrt{\epsilon_r} = 3 \times 10^8/$ $(2 \times 10^9 \times \sqrt{4.4}) = 0.0715$ m (i.e., 7.15 cm). Due to end effects, the value of b will need to be shortened by length 2Δ where $\Delta \approx 0.42h(\epsilon_r + 0.3)/(\epsilon_r - 0.258)$. For the current design $\Delta = 1.4\,\text{mm}$ and so $b = 6.87\,\text{cm}$ (we choose $a = 6\,\text{cm}$ to ensure that this represents the shortest side of the patch). The feed will need to be 50 Ω and this will require a microstrip of width 5.7 mm. The impedance at the edge of the patch is given by $60\lambda_o/a = 300\,\Omega$. Consequently, the feed will need to penetrate a distance d into the patch, where d satisfies $50 = 300\cos^2[\pi d/6.87]$ (i.e., $d = 2.52$ cm). Since the microstrip penetrates into the patch, this will require a gap (i.e., no metal) between the microstrip feed and the patch itself. A gap equal to the substrate thickness is usually sufficient.

EXERCISES

(1) A short dipole (3 m in length) is required to operate at 10 MHz (the dipole is made of aluminium rod with $\sigma = 4 \times 10^7\,\mathrm{S/m}$). Find the diameter of rod that ensures a 20 kHz bandwidth. Calculate the antenna efficiency and find a network that would match this antenna to 50 Ω coaxial cable.

(2) A small circular loop (1 m in diameter) is required to operate at 10 MHz (the loop is made of aluminium tube with $\sigma = 4 \times 10^7\,\mathrm{S/m}$). Find the diameter of tube that ensures a 3kHz bandwidth. Calculate the antenna efficiency and find the value of capacitance that would be needed to resonate the loop.

(3) Given an antenna has radiation resistance R_{rad}, find the relationship between its effective length $\mathbf{h}_{eff}$ and its directivity D.

(4) A half-wave dipole can be modelled by three end-on ideal dipoles of length $\lambda/6$. The central dipole will carry the feed current I_m and the other two dipoles the current $I_m/2$. Assuming the dipole to have radiation resistance of 73 Ω, use this simple model to calculate the directivity of the antenna.

(5) A small rectangular loop antenna can be modelled as four ideal dipoles. Use this model to verify the expression for the vector potential of a small loop antenna.

(6) Find an expression for the directivity of two collinear short dipoles a distance $\lambda/4$ apart and carrying currents that are 90° out of phase. (You may assume the approximation $\cos x \approx 1 - 4x^2/\pi^2$ for $|x| < \pi/2$.)

(7) A plane wave is normally incident upon a screen with a circular aperture. Estimate the maximum directivity of the radiation that escapes through the aperture. Calculate the radiation field of the aperture.

(8) Design a patch antenna for 2.5 GHz that has an input impedance of 50 Ω. The antenna should use a 3 mm FR4 substrate ($\epsilon_r = 4.4$) and have a microstrip feedline.

SOURCES

Balanis, C. A. 1997. *Antenna Theory: Analysis and Design* (2nd edn). New York: John Wiley.

Collin, R. E. 1985. *Antennas and Radio Wave Propagation*. New York: McGraw-Hill.

DuHamel, R. H. and Chadwick, G. G. 1984. Frequency independent antennas. In *Antenna Engineering Handbook* (2nd edn). R. C. Johnson and H. Jasik (eds.). New York: McGraw-Hill.

Hall, G. J. (ed.). 1988. *The ARRL Antenna Book*. Newark, CT: American Radio Relay League.

Kuecken, J. A. 1996. *Antennas and Transmission Lines*. Mississippi. MFJ Publishing.

Krauss, J. D. 1988. *Antennas* (2nd edn). New York: McGraw-Hill.

Miller, E. K., Medgyesi-Mitschang, L. and Newman, E. H. (eds.). 1991. *Computational Electromagnetics: Frequency-domain Method of Moments*. Piscataway, NJ: IEEE Press.

Pozar, D. M. and Schaubert, D. H. (eds.). 1995. *Microstrip Antennas*: *The Analysis and Design of Microstrip Antennas and Arrays*. Piscataway, NJ: IEEE Press.

Pozar, D. M. 2001. *Microwave and RF Design of Wireless Systems*. New York: John Wiley.

Rumsey, V. H. 1966. *Frequency Independent Antennas*. New York: Academic Press.

Saunders, S. R. 1999. *Antennas and Propagation for Wireless Communication Systems*. Chichester: John Wiley.

Smith, A. A. 1998. *Radio Frequency Principles and Applications*. Piscataway, NJ: IEEE Press.

Statyman, W. L. and Thiele, G. A. 1981. *Antenna Theory and Design*. New York: John Wiley.

Straw, R. Dean (ed.). 1999. *The ARRL Handbook* (77th edn). Newark, CT: American Radio Relay League.

Tong, D. A. 1974. The normal-mode helical aerial. In *Radio Communication*, July 1974. Radio Society of Great Britain.

Wheeler, H. A. 1947. Fundamental limitations of small antennas. In *Proceedings of the IRE*. December 1947, pp. 1479–1484.

11 Propagation

Propagation is the process whereby the electromagnetic radiation from a transmitting antenna reaches a receiving antenna. We have already seen that the wave amplitude reduces as the inverse of distance from the transmitter and that there can be additional attenuation if the medium has non-zero conductivity (or other loss mechanisms). Up to now we have assumed that there is a line-of-sight path between transmitter and receiver, but communications are often required when there is no line-of-sight. Reflections can provide a possible mechanism, but this requires there to be suitably positioned reflecting surfaces. There are, however, other mechanisms (refraction and diffraction) that have the potential to provide non-line-of-sight propagation without reflecting surfaces.

In most propagation scenarios, there exist a variety of propagation paths and the received signal can be a complex convolution of the original transmitted signal. As a consequence, the received signal can suffer from phenomena such as fading and delay spread. It is clear that RF system performance can be heavily affected by the available propagation mechanisms and their study is the purpose of the present chapter.

11.1 Reciprocity theorem

Given an electromagnetic field $(\mathbf{H}_A, \mathbf{E}_A)$ arising from current distribution $\mathbf{J}_A$ and another field $(\mathbf{H}_B, \mathbf{E}_B)$ arising from current distribution $\mathbf{J}_B$, we will have (from Maxwell's equations)

$$\nabla \times \mathbf{H}_A = j\omega\epsilon\mathbf{E}_A + \mathbf{J}_A \tag{11.1}$$

$$\nabla \times \mathbf{E}_A = -j\omega\mu\mathbf{H}_A \tag{11.2}$$

$$\nabla \times \mathbf{H}_B = j\omega\epsilon\mathbf{E}_B + \mathbf{J}_B \tag{11.3}$$

$$\nabla \times \mathbf{E}_B = -j\omega\mu\mathbf{H}_B. \tag{11.4}$$

On noting the identity

$$\nabla \cdot (\mathbf{X} \times \mathbf{Y}) = \mathbf{Y} \cdot \nabla \times \mathbf{X} - \mathbf{X} \cdot \nabla \times \mathbf{Y} \tag{11.5}$$

we obtain that

$$\nabla \cdot (\mathbf{E_A} \times \mathbf{H_B} - \mathbf{E_B} \times \mathbf{H_A}) = \mathbf{H_B} \cdot \nabla \times \mathbf{E_A} - \mathbf{E_A} \cdot \nabla \times \mathbf{H_B}$$
$$- \mathbf{H_A} \cdot \nabla \times \mathbf{E_B} + \mathbf{E_B} \cdot \nabla \times \mathbf{H_A} \tag{11.6}$$

and, substituting the Maxwell equations,

$$\nabla \cdot (\mathbf{E_A} \times \mathbf{H_B} - \mathbf{E_B} \times \mathbf{H_A}) = \mathbf{J_A} \cdot \mathbf{E_B} - \mathbf{J_B} \cdot \mathbf{E_A}. \tag{11.7}$$

Integrating over volume V (surface S) and applying the divergence theorem ($\int_V \nabla \cdot \mathbf{F} \, dV = \int_S \mathbf{F} \cdot \mathbf{n} \, dS$, where $\mathbf{n}$ is the unit normal on the surface S), we obtain

$$\int_S (\mathbf{E_A} \times \mathbf{H_B} - \mathbf{E_B} \times \mathbf{H_A}) \cdot \mathbf{n} \, dS = \int_V (\mathbf{J_A} \cdot \mathbf{E_B} - \mathbf{J_B} \cdot \mathbf{E_A}) \, dV. \tag{11.8}$$

We will consider the special case where the current sources are of bounded extent. For any surface S that surrounds all sources, the right-hand side of Equation 11.8 will yield the same value. If S is a sphere of sufficiently large radius, the fields will behave in a wave-like fashion on this surface (propagation in the radial direction). As a consequence, the integrand on the left-hand side will be zero and hence the integral itself. This will mean that the right-hand side will also be zero for any volume V that contains all the sources. As a consequence,

$$\int_S (\mathbf{E_A} \times \mathbf{H_B} - \mathbf{E_B} \times \mathbf{H_A}) \cdot \mathbf{n} \, dS = 0 \tag{11.9}$$

for any surface S that surrounds all sources and

$$\int_V \mathbf{J_A} \cdot \mathbf{E_B} \, dV = \int_V \mathbf{J_B} \cdot \mathbf{E_A} \, dV \tag{11.10}$$

for a volume V that contains all sources. Equations 11.9 and 11.10 represent the most general form of the *Lorentz reciprocity theorem*.

A variant of the theorem (Monteath) arises when the volume V only contains the sources of field A. In this case,

$$\int_V \mathbf{J_A} \cdot \mathbf{E_B} \, dV = \int_S (\mathbf{E_A} \times \mathbf{H_B} - \mathbf{E_B} \times \mathbf{H_A}) \cdot \mathbf{n} \, dS. \tag{11.11}$$

Example   A quarter-wave monopole is placed a distance d from the closed end of a waveguide and at the centre of the floor. The monopole generates a TE_{10} mode at a frequency ω for which this is the only propagating mode. If the waveguide has height b and width a, calculate the strength of the mode in terms of the current at the base of the monopole. Show that maximum amplitude occurs when $d = \lambda_{10}/4 = \pi/2\beta_{10}$ (β_{10} is the propagation constant for the TE_{10} mode) and find the input resistance at the monopole base.

We can use the idea of images to replace the closed end of the waveguide by an *image* monopole (as shown in Figure 11.1). Let $(\mathbf{H_A}, \mathbf{E_A})$ be the field caused by the monopoles

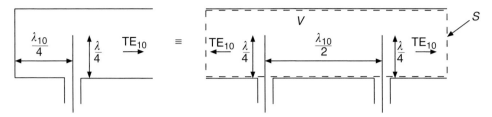

Figure 11.1   The excitation of a waveguide.

and let $(\mathbf{H_B}, \mathbf{E_B})$ represent a TE_{10} mode with $\mathbf{E_B} = \hat{y}\sin[\pi x/a]\exp(-j\beta_{10}z)$ (note that the $\hat{x}$ component of $\mathbf{H_B}$ is given by $-E_B/Z_{10}$ where $Z_{10} = \omega\mu/\beta_{10}$ is the wave impedance of the TE_{10} mode). On the surfaces of the waveguide, the tangential components of both $\mathbf{E_A}$ and $\mathbf{E_B}$ will be zero. Furthermore, assuming the waveguide to be long enough for the evanescent modes to have died away, the field $\mathbf{E_A}$ will have the form $\mathbf{E_A} = \hat{y}C\sin[\pi x/a]\exp(-j\beta_{10}z)$ far to the right and $\mathbf{E_A} = \hat{y}C\sin[\pi x/a]\exp(+j\beta_{10}z)$ far to the left (C is the, as yet, unknown field magnitude and must be the same for both left and right components). It should be noted that, to the right, the $\hat{x}$ component of $\mathbf{H_A}$ is given by $-E_A/Z_{10}$ and, to the left, by E_A/Z_{10}. Let V be the inside volume of the waveguide (surface S), then

$$\int_V \mathbf{J_A} \cdot \mathbf{E_B}\, dV = I[\exp(-j\beta_{10}d) - \exp(+j\beta_{10}d)] \int_0^{\frac{\lambda}{4}} \cos\left(2\pi\frac{y}{\lambda}\right) dy$$

$$= \frac{\lambda I}{j\pi}\sin(\beta_{10}d) \tag{11.12}$$

on noting that the current distribution is given by

$$\mathbf{J_A} = I\cos\left(2\pi\frac{y}{\lambda}\right)\delta\left(x - \frac{a}{2}\right)[\delta(z - d) - \delta(z + d)]\hat{y} \quad \text{for} \ \ 0 \le y \le \frac{\lambda}{4}$$

(the image current is opposite in direction to the real current). Furthermore, noting that there are no contributions from the walls of the waveguide,

$$\int_S (\mathbf{E_A} \times \mathbf{H_B} - \mathbf{E_B} \times \mathbf{H_A}) \cdot \mathbf{n}\, dS = \int_{E_L + E_R} [(\mathbf{n} \times \mathbf{E_A}) \cdot \mathbf{H_B} - (\mathbf{n} \times \mathbf{E_B}) \cdot \mathbf{H_A}]\, dS, \tag{11.13}$$

where E_L and E_R refer to the left and right ends of the waveguide. The contributions at end E_R will cancel and, as a consequence,

$$\int_S (\mathbf{E_A} \times \mathbf{H_B} - \mathbf{E_B} \times \mathbf{H_A}) \cdot \mathbf{n}\, dS = -2\int_{E_L} \frac{E_A E_B}{Z_{10}}\, dS$$

$$= -C\frac{2}{Z_{10}}\int_0^b\int_0^a \sin^2\left(\frac{\pi}{a}x\right) dx\, dy = -C\frac{ab}{Z_{10}}. \tag{11.14}$$

Equation 11.11, together with 11.12 and 11.14, will now yield $C = j I\lambda Z_{10}\sin(\beta_{10}d)/\pi ab$. It is clear from this that C will take its maximum value when $d = \lambda_{10}/4$. The power issuing through surface E_R will be given by $P = \frac{1}{2}\Re\{\int_{E_R}(\mathbf{E}_A \times \mathbf{H}_A) \cdot \mathbf{n}\,dS\} = ab|C|^2/4Z_{10}$ and will have the maximum value $Z_{10}\lambda^2 I^2/4\pi^2 ab$. The input resistance is given by $R = 2P/I^2$ and will have the value $Z_{10}\lambda^2/2\pi^2 ab$ when the power transfer is maximum.

11.2   Some consequences of reciprocity

Consider two antennas (A and B) and let the field $(\mathbf{H}_A, \mathbf{E}_A)$ result when A is driven and B is open circuit. Furthermore, let field $(\mathbf{H}_B, \mathbf{E}_B)$ result when B is driven and A is open circuit. The electric field will be zero inside the metal of the antennas and, as a consequence, the only contribution to the volume integrals of Equation 11.10 will come from the region of the antenna feeds. We can model a driven feed as a cylindrical region $S \times L$ (L represents the length and S the cross-section) across which there is a voltage drop $V = -\int_L \mathbf{E} \cdot d\mathbf{r}$ and through which there is a current flow $I = \int_S \mathbf{J} \cdot d\mathbf{S}$. We note from Equation 11.10 that

$$\int_{S_A} \mathbf{J}_A \cdot d\mathbf{S} \int_{L_A} \mathbf{E}_B \cdot d\mathbf{r} = \int_{S_B} \mathbf{J}_B \cdot d\mathbf{S} \int_{L_B} \mathbf{E}_A \cdot d\mathbf{r}, \qquad (11.15)$$

where $S_A \times L_A$ is the feed of A and $S_B \times L_B$ is the feed of B. Consequently, if V_{AB} is the open-circuit voltage across the terminals of A due to a current I_B in the feed of B and V_{BA} is the voltage drop across the terminals of B due to a current I_A in the feed of A,

$$I_A V_{AB} = I_B V_{BA} \qquad (11.16)$$

and, for the same drive current ($I_A = I_B$), we obtain

$$V_{AB} = V_{BA} \qquad (11.17)$$

which is the standard reciprocity result for antennas (see Chapter 1).

Now consider a surface S that separates antennas A and B (Figure 11.2). If this surface stretches to infinity, it can be closed by a surface on which the integrand of Equation 11.11 vanishes. As a consequence, Equation 11.11 will imply

$$-I_A V_{AB} = \int_S (\mathbf{E}_A \times \mathbf{H}_B - \mathbf{E}_B \times \mathbf{H}_A) \cdot \hat{z}\,dS. \qquad (11.18)$$

The major contribution to the integral will come from points close to the intersection of S with the line-of-sight between antennas A and B. As we move away from this intersection, the phase of the fields will increase and hence cause oscillations in the integrand. Furthermore, the rate of this oscillation will rapidly increases with distance away from the intersection. During the integration process, there will be cancellation between the positive and negative excursions of the oscillations. The point of intersection, however, is located at a phase minimum and points in its vicinity will suffer least

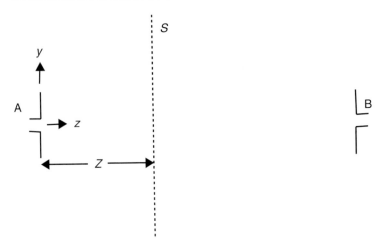

Figure 11.2   Interaction of two antennas.

from these cancellations. As a consequence, the integral can be accurately evaluated by replacing the integrand with an approximation that is valid close to the point of inter-section. It is assumed that antennas A and B are sufficiently removed from S for a plane wave approximation to be valid near the point of intersection, so that we can assume $\mathbf{H}_A \approx (\hat{z} \times \mathbf{E}_A)/\eta$ and $\mathbf{H}_B \approx -(\hat{z} \times \mathbf{E}_B)/\eta$. Equation 11.18 will now yield

$$I_A V_{AB} \approx 2 \int_S \frac{\mathbf{E}_A \cdot \mathbf{E}_B}{\eta} \, dS. \tag{11.19}$$

The field of antenna A will have the form

$$\mathbf{E}_A = \frac{j\omega\mu I_A}{4\pi} \mathbf{h}_{\text{eff}}^A \frac{\exp(-j\beta r)}{r}, \tag{11.20}$$

where $\mathbf{h}_{\text{eff}}^A$ is the effective length of antenna A. Furthermore, in the vicinity of A, the field of antenna B will behave as a plane wave with

$$\mathbf{E}_B = \mathbf{E}_B^A \exp(j\beta z), \tag{11.21}$$

where $\mathbf{E}_B^A$ is the value of $\mathbf{E}_B$ at the location of A. Consequently, if S is close enough to antenna A, we can substitute Equations 11.20 and 11.21 into Equation 11.19 to yield

$$V_{AB} \approx \frac{j\omega\mu}{2\pi\eta} \mathbf{h}_{\text{eff}}^A \cdot \mathbf{E}_B^A \int_{-\infty}^{\infty} \int_{-\infty}^{\infty} \frac{\exp[-j\beta(r - Z)]}{r} \, dx \, dy. \tag{11.22}$$

Since the major contribution to the integral will result from relatively small values of x and y, we can use the approximation $r = \sqrt{x^2 + y^2 + Z^2} \approx Z + (x^2 + y^2)/2Z$ to obtain

$$V_{AB} \approx \frac{j\omega\mu}{2\pi\eta} \mathbf{h}_{\text{eff}}^A \cdot \mathbf{E}_B^A \int_{-\infty}^{\infty} \int_{-\infty}^{\infty} \frac{\exp\left[-\frac{j\beta}{2Z}(x^2 + y^2)\right]}{Z} \, dx \, dy. \tag{11.23}$$

The integral is now the double application of the well-known mathematical result

$$\int_{-\infty}^{\infty} \exp(-j\alpha x^2)\, dx = \sqrt{\frac{\pi}{j\alpha}} \tag{11.24}$$

and so Equation 11.23 yields

$$V_{AB} = \mathbf{h}_{\text{eff}}^{A} \cdot \mathbf{E}_{B}^{A} \tag{11.25}$$

(a result that we have merely assumed up to this point). In terms of the effective length $\mathbf{h}_{\text{eff}}^{B}$ of antenna B

$$V_{AB} = \frac{j\omega\mu\, \mathbf{h}_{\text{eff}}^{A} \cdot \mathbf{h}_{\text{eff}}^{B}\, I_B}{4\pi} \frac{\exp(-j\beta R_{AB})}{R_{AB}}, \tag{11.26}$$

where R_{AB} is the distance between the antennas and I_B is the current in antenna B. As a consequence, the mutual impedance between the antennas is given by

$$Z_{AB} = \frac{j\omega\mu\, \mathbf{h}_{\text{eff}}^{A} \cdot \mathbf{h}_{\text{eff}}^{B}}{4\pi} \frac{\exp(-j\beta R_{AB})}{R_{AB}}. \tag{11.27}$$

11.3 Line-of-sight propagation and reflections

Consider the line-of-sight communications between two stations (A and B), a distance r_{AB} apart (Figure 11.3). If their respective antennas have gains G_A and G_B towards each other, the definition of gain yields

$$G_A G_B = \frac{\frac{r_{AB}^2}{2\eta}|\mathbf{E}_A^B|^2}{P_A/4\pi} \frac{\frac{r_{AB}^2}{2\eta}|\mathbf{E}_B^A|^2}{P_B/4\pi}, \tag{11.28}$$

where P_A and P_B are the total powers transmitted by A and B, respectively, when in transmit mode, $\mathbf{E}_A^B$ is the field of A evaluated at B and $\mathbf{E}_B^A$ is the field of B evaluated at A. Consequently, noting Expression 11.20 for $\mathbf{E}_A$,

$$P_B G_A G_B = \frac{(4\pi)^2 r_{AB}^4}{4\eta^2} \frac{|\mathbf{E}_B^A|^2}{P_A} \left(\frac{\omega\mu I_A |\mathbf{h}_{\text{eff}}^{A}|}{4\pi r_{AB}}\right)^2, \tag{11.29}$$

where I_A is the current that drives antenna A in transmit mode (note that $P_A = R_A I_A^2/2$, where R_A is the radiation resistance of antenna A). Now consider the situation where

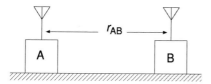

Figure 11.3   Communication system with two stations.

the antennas are polarisation matched, then

$$|\mathbf{h}^A_{\text{eff}}||\mathbf{E}^A_B| = |\mathbf{h}^A_{\text{eff}} \cdot \mathbf{E}^A_B| = V_A, \qquad (11.30)$$

where V_A is the open-circuit voltage in antenna A when B is in the transmit role. In this case, Equation 11.29 yields

$$P_B G_A G_B = \frac{r^2_{AB} \omega^2 \mu^2 V^2_A}{2\eta^2 R_A}. \qquad (11.31)$$

Let P_R be the received power at station A when B transmits, then $P_R = V^2_A/8R_A$ (maximum available power). Consequently, since $\omega\mu/\eta = 2\pi/\lambda$, Equation 11.31 reduces to the Friis equation

$$P_R = P_B G_A G_B \left(\frac{\lambda}{4\pi r_{AB}}\right)^2. \qquad (11.32)$$

From the above derivation, it is clear that the Friis equation will only apply when the antennas are *polarisation matched*. As a consequence, the *polarisation efficiency*

$$\eta_P = \frac{|\mathbf{h}^A_{\text{eff}} \cdot \mathbf{E}^A_B|^2}{|\mathbf{h}^A_{\text{eff}}|^2|\mathbf{E}^A_B|^2} \qquad (11.33)$$

will need to be taken into account if the antennas have different polarisations.

The Friis equation assumes that all field effects have been incorporated into the antenna gain patterns. For complex environments, however, such incorporation can be extremely difficult. Consequently, we will need to make a careful choice as to what aspects of the environment we treat as part of the antenna and what we treat as part of the *propagation*. Since any one antenna could be required to operate in a variety of environments, it is quite common to characterise the antenna when operating in free space and to treat all other effects as part of propagation. This means that we must now consider the ground reflections as propagation (as illustrated in Figure 11.4) and the Friis equation will need modification. The reflected field will be the weaker since it has travelled further (field strength still falls away in proportion to the inverse of the distance travelled) and has also suffered losses during ground reflection (the reflection coefficients of Chapter 9 can be used to estimate this loss). In addition, the field orientation at reception will be different from that transmitted. Most importantly, however, the extra distance travelled by the reflected wave will result in a phase delay

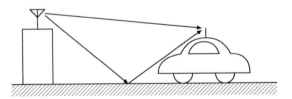

Figure 11.4   Propagation with ground reflections.

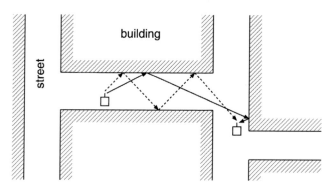

Figure 11.5   Propagation through an urban environment.

when compared to the directly propagating wave. When the fields combine, this will not necessarily be constructive and there are situations where the fields can almost cancel. This effect is quite pronounced in the case of mobile communications where the signal can experience *flutter* as the user moves between regions of constructive and destructive field combinations. Such effects can get worse in urban environments where the paths can be more tortuous and numerous (see Figure 11.5). For all paths, the signal strength will fall off as the inverse of distance travelled and the formulas of Chapter 9 can be used to calculate reflection losses. In the calculation of losses, values will be required for the material properties of the reflecting media and these are typically $\sigma \approx 0.01$ for agricultural land, and $\sigma = 0.02$ for rock. The corresponding values of permittivity are 10 and 5, respectively (the permeability can be taken to be that of free space).

Example   Two radio stations communicate at a frequency of 500 MHz by means of vertical half-wave dipole antennas. They sit on top of mountains of height h above sea level and are a distance l apart. If the valley between them contains a sea loch (see Figure 11.6), investigate the voltage induced in one antenna when the other is excited.

The voltage V_D in the terminals of the receive antenna caused by direct propagation will be $E_D h_{eff}$, where E_D is the vertical component of the direct field and h_{eff} ($\approx 0.32\lambda$, where $\lambda = 0.6$ m) is the effective length of a dipole for radiation that is orthogonal to its axis. As a consequence

$$V_D = \frac{j\omega\mu I h_{eff}^2}{4\pi} \frac{\exp(-j\beta l)}{l}, \tag{11.34}$$

where I is the current in the transmit antenna. The indirect wave will travel a distance $s = \sqrt{4h^2 + l^2}$ and will arrive at the receiver an angle $\theta = \arctan(2h/l)$ below the horizontal. For a frequency of 500 MHz, we can use the large conductivity limit $\eta = (1 + j)\sqrt{\omega\mu/2\sigma}$ to describe the seawater ($\sigma = 5$ S/m) and from which $\eta \approx 20(1 + j)$ under the assumption that $\mu = \mu_0$. As a consequence, the relative impedance η_r can

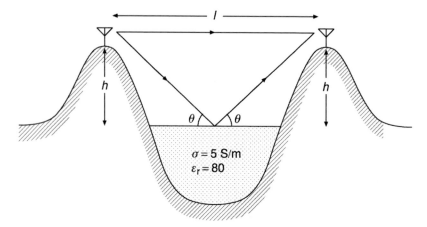

Figure 11.6 Propagation over a stretch of water.

be approximated by $0.053(1 + j)$. For such small values of impedance, the reflection coefficient R_V can be approximated by

$$R_V \approx \frac{\sin\theta - \eta_r}{\sin\theta + \eta_r} \tag{11.35}$$

from which it can be seen that there is a change from $R_V \approx -1$ at very small values of θ to $R_V \approx 1$ at higher values (the angle of transition is known as the *Brewster angle*). We will assume that angle θ is small enough for the dipole effective length in the direction of reflected radiation to be approximately the same as that for direct radiation ($h_{eff} \approx 0.32\lambda$). Since we have vertical polarisation, the voltage V_R due to the reflected signal is given by

$$V_R = \frac{j\omega\mu I h_{eff}^2}{4\pi} \frac{\exp(-j\beta s)}{s} R_V, \tag{11.36}$$

where R_V represents the change in field magnitude, and phase, caused by the reflection. Consequently, the open circuit voltage V in the receive antenna is given by

$$V = V_D + V_R \tag{11.37}$$

$$= \frac{j\omega\mu I h_{eff}^2}{4\pi} \left[\frac{\exp(-j\beta l)}{l} + R_V \frac{\exp(-j\beta s)}{s} \right] \tag{11.38}$$

$$\approx \frac{j\omega\mu I h_{eff}^2}{4\pi} \frac{\exp(-j\beta l)}{l} \left[1 + R_V \exp\left(-\frac{2j\beta h^2}{l} \right) \right] \tag{11.39}$$

on noting that, for small θ, $s \approx l + (2h^2/l)$. For very small values of θ the direct and reflected signals will cancel ($V \approx 0$) since $R_V \approx -1$. There will, however, still be a small amount of power that reaches the receiver through what is known as *ground wave* propagation, but the field in this case will fall off as l^{-2}. At larger values of θ (well above the Brewster angle) we will have $R_V \approx 1$ and there can be constructive interference between direct and reflected signals (an enhanced voltage V in the antenna). Even at these larger values of θ, however, there can still be problems since there will be strong destructive interference if $4h^2/l\lambda = 2m + 1$ (m any integer). In the case that one of the stations is moving (an aircraft, for example), this can result in the signal flutter already mentioned.

11.4 Diffraction

If the receiver and transmitter have no direct path, or path by reflection, they can still communicate through a process known as *diffraction*. A small amount of energy will be diffracted into the *shadow* region of an object and this will often be sufficient for communication purposes. The scenario of Figure 11.7 shows propagation over a building by means of such a process. The diffraction phenomena can be explained through *Huygen's principle*:

Each point on the wavefront of a general wave can be considered as the source of a secondary spherical wave. A subsequent wavefront of the general wave can then be constructed as the envelope of secondary wavefronts.

The principle can be applied to diffraction over a screen, as illustrated in Figure 11.8. For strictly plane waves, the non-plane components of the spherical waves will cancel to leave a plane wave. At the screen, however, there are no cancelling components below the screen edge and this leaves some spherical components that propagate into the shadow region. Huygen's principle finds its mathematical expression in the integral formulations of Section 11.1. Consequently, we can modify the derivation of Equation 11.25 to take account of propagation that is partially obscured by a screen (see Figure 11.9). The screen is orthogonal to the line joining the transmitter and receiver and distance Z from antenna A. It is of infinite extent in both x-directions and the negative y-direction. In the positive y-direction, however, the screen has a finite height

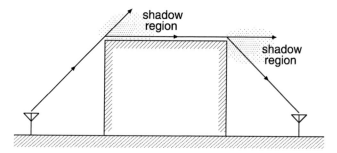

Figure 11.7   Diffractive propagation over a building.

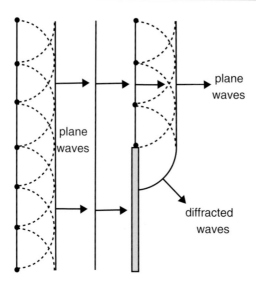

Figure 11.8 Huygen's principle and diffraction over a screen.

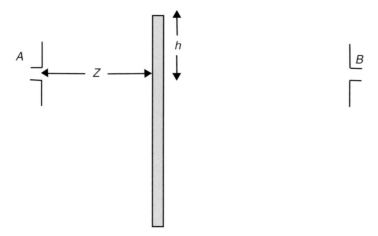

Figure 11.9 Propagation obscured by a screen.

h. Expression 11.23 is still applicable, providing that the y integral is limited to the region from h to ∞ (due to the blocking action of the screen). As a consequence,

$$V_{AB} = \frac{j\omega\mu}{2\pi\eta} \mathbf{h}_{eff}^A \cdot \mathbf{E}_B^A \int_h^\infty \int_{-\infty}^\infty \frac{\exp\left[-\frac{j\beta}{2Z}(x^2 + y^2)\right]}{Z} \, dx \, dy. \qquad (11.40)$$

We can perform the x integral as before and rescale the y integral to obtain

$$V_{AB} = \mathbf{h}_{eff}^A \cdot \mathbf{E}_B^A \sqrt{\frac{j}{\pi}} \int_{h\sqrt{\frac{\beta}{2Z}}}^\infty \exp\left(-jt^2\right) dt, \qquad (11.41)$$

where $t = y\sqrt{\beta/2Z}$. If we assume antenna A to be located in the Fresnel zone of the screen ($Z < h^2/\lambda$), we will have a value of $h\sqrt{\beta/2Z}$ that is relatively large. Consequently, since $\int_x^\infty \exp(-jt^2)\,dt \approx \exp(-jx^2)/2jx$ for large x, we obtain

$$V_{AB} = \mathbf{h}_{\text{eff}}^A \cdot \mathbf{E}_B^A \frac{\sqrt{\lambda Z}}{2\pi h\sqrt{j}} \exp\left(-jh^2\frac{\pi}{\lambda Z}\right). \qquad (11.42)$$

In decibels, the additional attenuation that is caused by the introduction of a screen will be

$$L = 20\log_{10}\left(\frac{2\pi h}{\sqrt{\lambda Z}}\right) \qquad (11.43)$$

provided that $h > \sqrt{\lambda Z}$ (note that Z should be the distance of the antenna that is closest to the screen).

11.5   Refraction

In addition to the mechanisms of diffraction and reflection, energy can also propagate in a non-line-of-sight fashion through a process known as *refraction*. This process, however, requires the properties of the propagation medium to be spatially varying. We have already seen that a plane wave can be reflected if it is intercepted by a material with different electrical properties. Except where this material is perfectly conducting, however, some of the incident energy will propagate into the new material. The speed of this propagation can be vastly different and cause the distance between wave fronts (surfaces in which the field values are constant) to change. We require the wavefronts to be continuous across the material interface and this will necessitate a change in propagation direction (see Figure 11.10). If c_A is the phase speed in medium A and c_B is that in medium B, continuity of the wave fronts will require

$$\frac{\sin\theta_A}{\sin\theta_B} = \frac{c_A}{c_B}, \qquad (11.44)$$

where θ_A is the angle between the incident wavefront and the interface and θ_B is the angle between the transmitted wavefront and the interface. This expression is known as *Snell's law* and quantifies the change in direction due to refraction. Snell's law will still hold if the medium is continuously varying. The plane interface is now replaced by a series of $c = $ constant surfaces with θ the angle between the propagation direction and the normal to the surface the wave is crossing. Consider propagation upwards from ground level through a medium that is horizontally stratified (c depends on z alone for Figure 11.11). The propagation direction will continuously vary according to

$$\frac{\sin\theta}{\sin\theta_A} = \frac{c}{c_A}, \qquad (11.45)$$

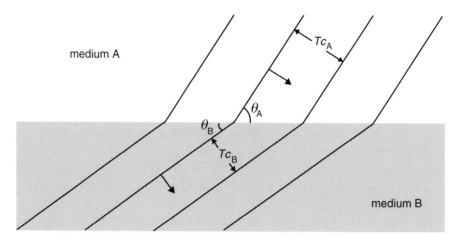

Figure 11.10   The refraction of a plane wave at a plane interface (T is the time between wave fronts).

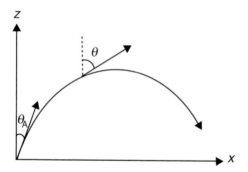

Figure 11.11   The direction of propagation with refraction by a continuously varying medium.

where θ is the angle of this direction to the vertical. The speed of propagation c is related to that of free space c_0 through $c = c_0/N$, where $N = \sqrt{\epsilon_r \mu_r}$ is known as the *refractive index*. From Equation 11.45 it will be noted that the propagation paths will bend towards a region of higher refractive index. The simplest variation of refractive index is one in which the square of the refractive index is linearly dependent upon height (taken to be the z coordinate)

$$N^2 = 1 + kz. \tag{11.46}$$

In this case, Equation 11.45 reduces to

$$\frac{\sin \theta}{\sin \theta_A} = \frac{1}{\sqrt{1 + kz}}. \tag{11.47}$$

Providing k is negative, the propagation path will bend towards the ground and will start returning to the ground when θ reaches $90°$, that is, at a height h_0 for which $1 + kh_0 = \sin^2 \theta_A$.

General refraction can be difficult to analyse, but some headway can be made in the case of short wavelengths (defined to be the circumstance in which the wavelength is much smaller than the scale L of material variations, i.e., $\beta \gg 1/L$). In this case, we assume a time-harmonic electric field of the form

$$\mathbf{E}(\mathbf{r}) = \mathbf{E}_0(\mathbf{r}) \exp[-j\beta_0 \phi(\mathbf{r})], \tag{11.48}$$

where $\beta_0 = \omega/c_0$ and c_0 is the speed of light in free space ($\mu_r = \epsilon_r = 1$). (Note that, for a plane wave in free space, $\mathbf{E}_0$ is constant and ϕ is given by $\phi = \hat{\mathbf{p}} \cdot \mathbf{r}$. In the case of a general radiating source located in a free space, $\mathbf{E}_0 = (j\beta I_m/4\pi\eta r)\mathbf{h}_{\text{eff}}$ and $\phi = |\mathbf{r}|$.) Eliminating $\mathbf{H}$ between the time-harmonic Maxwell equations, and assuming μ to be constant, we obtain

$$\nabla \times \nabla \times \mathbf{E} - \omega^2 \mu\epsilon \mathbf{E} = 0. \tag{11.49}$$

Then, noting the vector identity $\nabla \times \nabla \times \mathbf{E} = \nabla(\nabla \cdot \mathbf{E}) - \nabla^2 \mathbf{E}$ and that $\nabla \cdot (\epsilon\mathbf{E}) = 0$, we obtain

$$\nabla^2 \mathbf{E} + \nabla[(\nabla \ln N^2) \cdot \mathbf{E}] + \beta_0^2 N^2 \mathbf{E} = 0. \tag{11.50}$$

Substituting the trial solution 11.48 into Equation 11.50, and retaining only leading order terms (those in β_0 and β_0^2), we obtain

$$-j\beta_0\nabla^2\phi\mathbf{E}_0 - 2j\beta_0\nabla\phi \cdot \nabla\mathbf{E}_0 - \beta_0^2\nabla\phi \cdot \nabla\phi\mathbf{E}_0 - j\beta_0\nabla\ln(N^2) \cdot \mathbf{E}_0\nabla\phi + \beta_0^2 N^2 \mathbf{E}_0 = 0. \tag{11.51}$$

The β_0^2 term yields

$$\nabla\phi \cdot \nabla\phi - N^2 = 0 \tag{11.52}$$

and the β_0 term yields

$$-\nabla^2\phi\mathbf{E}_0 - 2\nabla\phi \cdot \nabla\mathbf{E}_0 - \nabla\ln(N^2) \cdot \mathbf{E}_0\nabla\phi = 0. \tag{11.53}$$

Starting at a field source (position $\mathbf{r} = \mathbf{r}_S$), it is possible to define a set of curves $\mathbf{r} = \mathbf{r}(g)$, known as *ray paths*, that are everywhere orthogonal to the wavefronts (the surfaces of constant phase ϕ that are shown in Figure 11.12). We define a curve parameter g such that

$$\frac{d\mathbf{r}}{dg} = \nabla\phi \tag{11.54}$$

and hence

$$\frac{d\mathbf{r}}{dg} \cdot \frac{d\mathbf{r}}{dg} = N^2. \tag{11.55}$$

It is obvious that $g = \int_{\mathbf{r}_s}^{\mathbf{r}} ds/N$, where the integral is taken along the ray path and s is the geometric distance along that path. The geometric distance provides an alternative

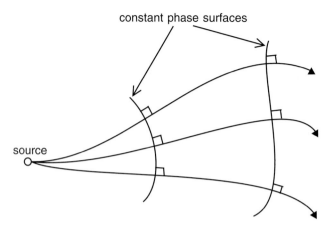

Figure 11.12   Propagation through a spatially varying medium.

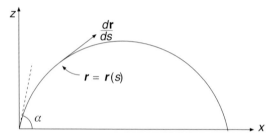

Figure 11.13   A ray path in a spatially varying medium.

parameterisation of the ray path. It will be noted that the quantity $p = \int_{\mathbf{r}_s}^{\mathbf{r}} N \, ds$, known as the *phase distance*, yields the value of ϕ at point $\mathbf{r}$ on the ray. Taking the derivative of Equation 11.55 along the ray path, we obtain

$$2\frac{d^2\mathbf{r}}{dg^2} \cdot \frac{d\mathbf{r}}{dg} = 2N\frac{d\mathbf{r}}{dg} \cdot \nabla N \tag{11.56}$$

which is satisfied when

$$\frac{d^2\mathbf{r}}{dg^2} = N\nabla N. \tag{11.57}$$

The above equations can be used to obtain a ray path once its source position and initial direction have been specified (Figure 11.13).

For a simple linear variation of N^2 (Equation 11.46), the ray equations yield

$$\frac{d^2x}{dg^2} = 0, \quad \frac{d^2y}{dg^2} = 0 \quad \text{and} \quad \frac{d^2z}{dg^2} = \frac{k}{2} \tag{11.58}$$

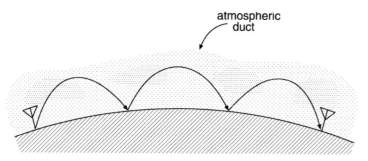

Figure 11.14 Ducting propagation in the atmosphere (refractive index decreasing with height).

and from which

$$x = g \cos \alpha, \quad y = 0 \quad \text{and} \quad z = \frac{kg^2}{4} + g \sin \alpha. \tag{11.59}$$

It will be noted that the ray paths return to the ground for k negative and so there is potential for propagation over an obstacle. Under normal atmospheric conditions, $N \approx 1.0003 - 0.000\,04z$ for z measured in kilometres and, although this behaviour will cause the propagation to curve back towards the ground, the curvature will be insufficient to counter that of the Earth. In general, however, the refractive index is related to the meteorological conditions through the Debye formula

$$N = 1 + \frac{7.76 \times 10^{-5}}{T} \left(P + 4810 \frac{e}{T} \right), \tag{11.60}$$

where T is the temperature (in kelvin), P is the atmospheric pressure (in millibars) and e is the water vapour pressure (in millibars). Atmospheric conditions can sometimes occur such that the refractive index decreases strongly with height, normally when there is a layer of warm air above a layer of cold air (a temperature inversion). Under these circumstances, refraction can cause the propagation to return to the ground over the horizon. Together with ground reflections, this can lead to *ducting* propagation that enables VHF communications over many thousands of kilometres (see Figure 11.14). The effect of the curvature of Earth can be included by employing the modified refractive index $M = N(1 + z/a)$, where $a = 6371$ km is the radius of the Earth. The coordinate z will now represent the altitude above the surface of the Earth and k will need to be less than $-0.000\,314$ (z is in kilometres) for the rays to return to the ground.

Several hundreds of kilometres up in the atmosphere, solar radiation generates a layer of ionised gas known as the *ionosphere* (see Figure 11.15). This plasma layer is dispersive and, in its lower regions, has a refractive index with negative vertical gradient. For frequencies below about 30 MHz, this layer can cause radio waves to be refracted back towards the ground and can thus provide reliable radio communications on a global scale. The refractive index of a plasma has the form $N = \sqrt{1 - f_p^2/f^2}$, where f is the wave frequency and f_p is the plasma frequency. Frequency f_p (in H$_z$) is related

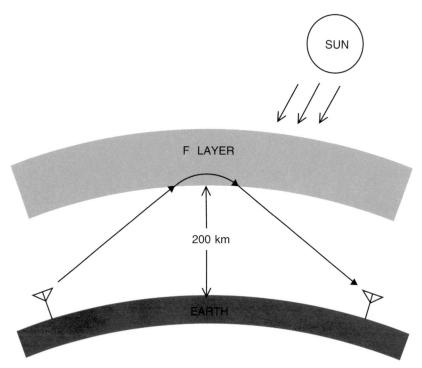

Figure 11.15   The ionospheric F layer.

to the plasma electron density N_e (electrons per cubic meter) through $f_p^2 = 80.5 N_e$. The plasma frequency of the main ionospheric layer (the F layer) can be modelled by

$$\frac{f_p^2}{f_m^2} = 1 - \frac{(z - h_m)^2}{y_m^2} \quad \text{for } |z - h_m| < y_m$$

$$= 0 \quad \text{otherwise}, \tag{11.61}$$

where h_m is the layer peak height (typically 300 km), y_m is its thickness (typically 100 km) and f_m is the peak plasma frequency (typically 10 MHz). From Equation 11.57 we obtain

$$\frac{\mathrm{d}}{\mathrm{d}s}\left(N\frac{\mathrm{d}x}{\mathrm{d}s}\right) = 0 \tag{11.62}$$

along a ray trajectory and, from which,

$$N(z)\frac{\mathrm{d}x}{\mathrm{d}s} = C \text{ (a constant).} \tag{11.63}$$

If θ is the angle between the z-direction and the ray, we obtain *Snell's law*

$$N(z)\sin\theta = N(0)\sin\theta_0, \tag{11.64}$$

where θ_0 is the value of θ at the start point of the ray ($z = 0$). For propagation that returns to the ground, there will be a height h at which the angle θ becomes $90°$. In the

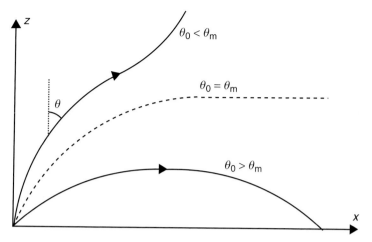

Figure 11.16　Ray paths in the ionosphere.

case of the above ionospheric model, Snell's law will reduce to

$$\frac{\sin^2 \theta - \sin^2 \theta_0}{\sin^2 \theta} = \frac{f_p^2}{f^2}$$

$$= \frac{f_m^2}{f^2}\left[1 - \frac{(z-h_m)^2}{y_m^2}\right] \quad \text{for } |z - h_m| < y_m. \quad (11.65)$$

Consequently, the height h at which the ray turns back towards the ground is given by

$$h = h_m - y_m\sqrt{1 - \frac{f^2}{f_m^2}\cos^2 \theta_0}. \quad (11.66)$$

From this expression, it is clear that there will be an initial angle $\theta_m = \cos^{-1}(f_m/f)$ for which lesser values of θ_0 will result in an imaginary value of square root and hence correspond to rays that cannot return to the ground (see Figure 11.16). From Equation 11.57 we also obtain that

$$N\frac{d}{ds}\left(N\frac{dx}{ds}\right) = \frac{1}{2}\frac{d(N^2)}{dz} \quad (11.67)$$

along a ray trajectory and, using Equation 11.63, we obtain

$$\frac{d^2z}{dx^2} = \frac{1}{2C^2}\frac{d(N^2)}{dz} \quad (11.68)$$

from which

$$\left(C\frac{dz}{dx}\right)^2 = N^2(z) + B, \quad (11.69)$$

where B is a constant of integration. Assuming $N(0) = 1$, we obtain that $C = \sin \theta_0$ and $B = -\sin^2 \theta_0$. As a consequence,

$$x(z) = \int_0^z \sqrt{\frac{\sin^2 \theta_0}{N^2(z) - \sin^2 \theta_0}} \, dz. \tag{11.70}$$

For propagation that returns to the ground, $x(h)$ will be the distance at which the ray starts turning downwards (h is given by Equation 11.66) and hence the total range of the propagation D will be $2x(h)$ (note that the ray path will be symmetric about its midpoint). For the above simple ionosphere

$$D = 2(h_m - y_m) \tan \theta_0 + 2 \int_{h_m - y_m}^{h} \frac{\sin \theta_0}{\sqrt{\cos^2 \theta_0 - \frac{f_m^2}{f^2} \left(1 - \frac{(z-h_m)^2}{y_m^2}\right)}} \, dz$$

$$= 2(h_m - y_m) \tan \theta_0$$

$$+ 2 \frac{y_m f \sin \theta_0}{f_m} \ln \left[\frac{y_m + \sqrt{y_m^2 - y_m^2 \left(1 - \frac{f^2}{f_m^2} \cos^2 \theta_0\right)}}{(h_m - h) + \sqrt{(h_m - h)^2 - y_m^2 \left(1 - \frac{f^2}{f_m^2} \cos^2 \theta_0\right)}} \right] \tag{11.71}$$

and, using Expression 11.66 for h, we obtain

$$D = 2(h_m - y_m) \tan \theta_0 + \frac{y_m f \sin \theta_0}{f_m} \ln \left(\frac{1 + \frac{f}{f_m} \cos \theta_0}{1 - \frac{f}{f_m} \cos \theta_0} \right). \tag{11.72}$$

It should be noted that not all ranges D can be reached by ionospheric propagation. In fact, there is a region (known as the *skip zone*) that is centred on the field source and within which no ray can land (i.e., there are no θ_0 corresponding to ranges D within this region). Another practical problem for ionospheric propagation is the possibility of power dissipation in the ionospheric medium itself. Such losses are relatively insignificant at night, but during the day they can be quite severe. During the day, there is an additional ionospheric layer at an altitude of about 70 km (the D layer) in which the loss mechanisms are extremely strong. These mechanisms can severely attenuate radio waves that pass through the D layer and can cause attenuation of the order of tens of decibels. Gigahertz frequency propagation in the lower atmosphere can also suffer considerable attenuation, but this is of the order of 0.1 dB/km for frequencies below about 60 GHz. Around 60 GHz, however, the resonance of molecular oxygen causes attenuation of the order of 10 dB/km and the attenuation remains at a fairly high level for frequencies above this.

Although ray Equations 11.57 provide us with a means of calculating the path of the radio wave, they do not provide information concerning the field strength. As with plane waves, the $\nabla \times \mathbf{E}$ Maxwell equation will imply

$$\mathbf{H} \approx \frac{1}{\eta}\frac{d\mathbf{r}}{ds} \times \mathbf{E} \tag{11.73}$$

and the $\nabla \cdot (\epsilon \mathbf{E})$ equation implies the condition

$$\mathbf{E} \cdot \frac{d\mathbf{r}}{ds} \approx 0 \tag{11.74}$$

(note that $d\mathbf{r}/ds$ is a unit vector in the propagation direction). Consequently, the Poynting vector $\mathbf{S}$ will be given by

$$\mathbf{S} = \frac{1}{2\eta}|\mathbf{E}|^2\frac{d\mathbf{r}}{ds} \tag{11.75}$$

and from which it is clear that energy will flow along ray paths. Since the ray paths will diverge away from a source, it is obvious that the field strength will fall away (the r^{-1} field dependence in free space). Consider a bundle of rays that intersect an area A_1 on a constant phase surface $\phi = \phi_1$ and an area A_2 on a later surface $\phi = \phi_2$. Since no energy flows across the rays of a bundle, the power flowing through area A_1 will be the same as the power flowing through area A_2. That is,

$$A_1\frac{|\mathbf{E}^1|^2}{\eta_1} = A_2\frac{|\mathbf{E}^2|^2}{\eta_2}, \tag{11.76}$$

where $\mathbf{E}^1$ is the field on $\phi = \phi_1$ and $\mathbf{E}^2$ is the field on $\phi = \phi_2$ (η_1 and η_2 are the corresponding impedances of the propagation medium at these surfaces). Expression 11.76 allows the field magnitude to be developed along a ray path and can be used to calculate the reduction in signal strength as a result of propagation between transmitter and receiver.

There still remains the question of the development of the field orientation. Taking the dot product of $\mathbf{E}_0^*$ with Equation 11.53, and noting that $\nabla\phi \cdot \mathbf{E}_0 = 0$, we obtain

$$-\nabla^2\phi\mathbf{E}_0 \cdot \mathbf{E}_0^* - \nabla\phi \cdot \nabla(\mathbf{E}_0^* \cdot \mathbf{E}_0) = 0 \tag{11.77}$$

and substituting back into Equation 11.53,

$$\nabla\phi \cdot \frac{\nabla(|\mathbf{E}_0|^2)}{|\mathbf{E}_0|^2}\mathbf{E}_0 - 2\nabla\phi \cdot \nabla\mathbf{E}_0 - 2\frac{\nabla N}{N} \cdot \mathbf{E}_0\nabla\phi = 0. \tag{11.78}$$

Since $d\mathbf{r}/dg = \nabla\phi$, we obtain

$$\frac{1}{|\mathbf{E}_0|}\frac{d}{dg}(|\mathbf{E}_0|)\mathbf{E}_0 - \frac{d\mathbf{E}_0}{dg} - \frac{\nabla N}{N} \cdot \mathbf{E}_0\frac{d\mathbf{r}}{dg} = 0 \tag{11.79}$$

along a ray path and, defining the *polarisation* vector by $\mathbf{P} = \mathbf{E}_0/|\mathbf{E}_0|$,

$$\frac{d\mathbf{P}}{dg} + \frac{\nabla N}{N} \cdot \mathbf{P}\frac{d\mathbf{r}}{dg} = 0 \qquad (11.80)$$

from which it is clear that the polarisation direction can change as a result of gradients in the refractive index. The polarisation vector, however, remains orthogonal to the direction of propagation.

11.6 Ground wave propagation

It has previously been noted that, for propagation that grazes the ground, direct and reflected waves will tend to cancel. This, however, does not mean that there will be no power propagated, merely that any such power will fall away as D^{-4} rather than the D^{-2} that is expected in free space (D is the distance from the transmitter). There will always be a small amount of power that is propagated by means of *ground waves* and such propagation does not require line of sight. Consider propagation over a hill (as depicted in Figure 11.17). This can be analysed by a slight modification to the analysis that led to Equation 11.42 (it is assumed that $R \gg Z$). The hill can be considered as a screen of height h, but the fields will now need to include the effects of ground reflection. Consequently, in deriving the ground wave version of Equation 11.42, the fields will need to be multiplied by factors of the form $1 + R_V$, where R_V is the plane wave reflection coefficient

$$R_V = \frac{\sin \alpha \, \eta_r^{-2} - \sqrt{\eta_r^{-2} - \cos^2 \alpha}}{\sin \alpha \, \eta_r^{-2} + \sqrt{\eta_r^{-2} - \cos^2 \alpha}}$$

$$\approx \frac{2\alpha}{\eta_r} - 1 \quad \text{for } \alpha \ll \eta_r \ll 1 \qquad (11.81)$$

with η_r the relative impedance of the ground and α the angle between the propagation direction and the ground. In the evaluation of Equation 11.40, the field values are only required for small values of y and hence for small values of α (note that $\alpha \approx (y - h)/Z$ and $\alpha \approx (y - h)/R$ for antennas A and B, respectively). Consequently,

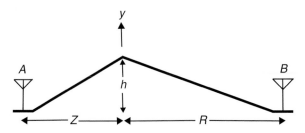

Figure 11.17 Ground wave propagation over a hill.

since η_r will be relatively small at low frequencies, the corresponding reflection coefficients (R_V^A and R_V^B) will satisfy $1 + R_V^A \approx 2(y-h)/Z\eta_r$ and $1 + R_V^B \approx 2(y-h)/R\eta_r$, respectively (providing that $\sin\alpha \ll \eta_r$). Equation 11.40 will therefore be replaced by

$$V_{AB} = \frac{j\omega\mu}{2\pi\eta}\, \mathbf{h}_{\mathrm{eff}}^A \cdot \mathbf{E}_B^A \int_h^\infty \int_{-\infty}^\infty \left(1 + R_V^A\right)\left(1 + R_V^B\right) \frac{\exp\left[-\frac{j\beta}{2Z}(x^2 + y^2)\right]}{Z}\, dx\, dy$$

$$\approx \frac{j}{\lambda}\, \mathbf{h}_{\mathrm{eff}}^A \cdot \mathbf{E}_B^A \int_h^\infty \int_{-\infty}^\infty \frac{4(y-h)^2}{ZR\eta_r^2}\, \frac{\exp\left[-\frac{j\beta}{2Z}(x^2 + y^2)\right]}{Z}\, dx\, dy. \tag{11.82}$$

The above integral can be approximated as

$$V_{AB} \approx \frac{2}{j\beta\eta_r^2 R}\mathbf{h}_{\mathrm{eff}}^A \cdot \mathbf{E}_B^A \tag{11.83}$$

and from which it will be noted that the ground wave is little affected by the hill, even though it prevents any line of sight. Furthermore, since signal strength increases as the frequency decreases, low frequency transmissions can provide reliable over-the-horizon communication (a fact that is exploited by users of the AM broadcast bands). In terms of effective length

$$V_{AB} \approx \frac{\eta_0 \mathbf{h}_{\mathrm{eff}}^A \cdot \mathbf{h}_{\mathrm{eff}}^B}{2\pi\eta_r^2 D^2}\, I_B\, \exp(-j\beta D), \tag{11.84}$$

where I_B is the current flowing into antenna B and D is the total ground distance between receiver and transmitter. (Note that effective lengths $\mathbf{h}_{\mathrm{eff}}^A$ and $\mathbf{h}_{\mathrm{eff}}^B$ are those appropriate to free space.)

Thus far our considerations have assumed a reflection coefficient that is appropriate to vertically polarised waves and hence to vertical antennas. For horizontal antennas, we need to use the reflection coefficient that is appropriate to horizontal polarisation. In this case, however, it turns out that the ground wave is negligible, a fact that emphasises the importance of vertical polarisation for ground wave communications.

11.7 Propagation by scattering

If we consider an electromagnetic wave that is normally incident upon a plane metallic plate (see Figure 11.18), we would expect there to be some direct reflections at the surface of the plate and some diffraction of energy into the shadow region behind the plate. It turns out, however, that there will be significant amounts of energy *scattered* in all directions to the fore of the plate. The incident wave will induce currents in the plate and these will radiate. Close to the centre of the plate, the off-perpendicular components of the radiated fields will cancel to form an outgoing plane wave. At the edges, however, the induced currents will cease and off perpendicular fields will no

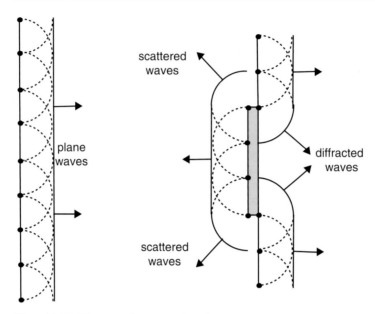

Figure 11.18   The scattering caused by a flat plate.

longer cancel. Radiation will therefore form in off-perpendicular directions. For an incident plane wave with magnetic field $\mathbf{H}_a$, we have already seen that there will be surface current density $\mathbf{J}_a = 2\mathbf{n} \times \mathbf{H}_a$ induced in the plate (see Section 10.7) and that this will cause an electric field of the form

$$\mathbf{E} = \frac{j\omega\mu \exp(-j\beta r)}{2\pi r}[\hat{\mathbf{r}}\hat{\mathbf{r}} \cdot (\mathbf{n} \times \mathbf{H}_a) - \mathbf{n} \times \mathbf{H}_a]\int_A \exp(j\beta\hat{\mathbf{r}} \cdot \mathbf{r}')\, dS', \qquad (11.85)$$

where $\hat{\mathbf{r}}$ is the unit vector in the radial direction. It is clear the there is considerable radiation in other than the normal direction $\mathbf{n}$.

Example   A square plate (sides of length $2a$) is illuminated by a normally incident plane wave. Calculate the scattered field.

We assume the plate to lie in the $z = 0$ plane, then the scattered field is given by

$$\mathbf{E} = \frac{j\omega\mu \exp(-j\beta r)}{2\pi r}(\hat{\mathbf{r}}\hat{\mathbf{r}} \cdot (\mathbf{n} \times \mathbf{H}_a) - \mathbf{n} \times \mathbf{H}_a)\int_{-a}^{a}\int_{-a}^{a} \exp(j\beta(\hat{r}_x x' + \hat{r}_y y'))\, dx' dy'$$

$$= \frac{j\omega\mu \exp(-j\beta r)}{2\pi r}\frac{\hat{\mathbf{r}}\hat{\mathbf{r}} \cdot \mathbf{E}_a - \mathbf{E}_a}{\eta}\frac{4\sin(\beta a\hat{r}_x)\sin(\beta a\hat{r}_y)}{\hat{r}_x\hat{r}_y\beta^2}, \qquad (11.86)$$

where $\mathbf{E}_a = \eta\mathbf{n} \times \mathbf{H}_a$ is the incident electric field, $\mathbf{n}$ is unit normal and $\hat{\mathbf{r}}$ is the unit vector in the direction of the observer. The scattered field is strongest in the direction of direct reflection ($\hat{\mathbf{r}} = \mathbf{n}$) and falls away to either side. Depending on the size of the plate, however, the field can exhibit several nulls. (This is illustrated in Figure 11.19 for the case of a square plate with sides of length 2λ.)

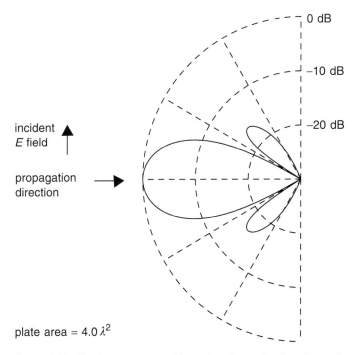

incident
E field

propagation
direction

plate area = $4.0\,\lambda^2$

Figure 11.19   Total power scattered by a plate (normalised on the maximum power).

Scattering does not necessarily require a metallic scatterer and any compact region of permittivity anomaly will also cause an incident electromagnetic wave to be scattered. Except for resonant structure, however, the scattering will be relatively weak. We will investigate the scattering caused by permittivity variations through a modified form of the reciprocity principle (see Monteath). In the derivation of the reciprocity theorem, the media associated with the separate electromagnetic fields $(\mathbf{H}_A, \mathbf{E}_A)$ and $(\mathbf{H}_B, \mathbf{E}_B)$ were assumed to have the same permittivity ($\epsilon_A = \epsilon_B = \epsilon$). If, however, there is a region V^+ over which these quantities differ ($\epsilon_A = \epsilon$ and $\epsilon_B = \epsilon + \delta\epsilon$), the reciprocity result will exhibit the modified form

$$\int_V \mathbf{J}_A \cdot \mathbf{E}_B \, dV - \int_V \mathbf{J}_B \cdot \mathbf{E}_A \, dV = -j\omega \int_{V^+} \delta\epsilon \mathbf{E}_B \cdot \mathbf{E}_A \, dV. \qquad (11.87)$$

Let $(\mathbf{H}_A, \mathbf{E}_A)$ be the field due to antenna A when driven by current I and antenna B open circuit. Furthermore, let $(\mathbf{H}_B, \mathbf{E}_B)$ be the field due to antenna B when driven by current I and antenna A open circuit (see Figure 11.20). The modified form of the reciprocity result will yield

$$I\delta V = -j\omega \int_{V^+} \delta\epsilon \, \mathbf{E}_B \cdot \mathbf{E}_A \, dV, \qquad (11.88)$$

where $\delta V = V_A - V_B$ with V_A the voltage in antenna A due to current I in antenna B and V_B the voltage in antenna B due to current I in antenna A. Effectively, δV is

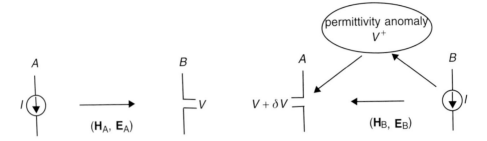

(a) Direct propagation between antennas

(b) Propagation augmented by scatter from a permittivity anomaly

Figure 11.20   Propagation by scatter from a permittivity anomaly.

the additional voltage in the antenna caused by scatter from the permittivity anomaly. We will assume that the anomaly is only a small perturbation to an otherwise uniform permittivity and hence use the unperturbed $\mathbf{E}_B$ when evaluating the integral in Equation 11.88. As a consequence

$$I \delta V = j \frac{\omega^3 \mu^2 \epsilon I^2 \mathbf{h}_{\text{eff}}^A \cdot \mathbf{h}_{\text{eff}}^B}{16\pi^2} \int_{V^+} \frac{\delta\epsilon}{\epsilon} \frac{\exp -j\beta(r_A + r_B)}{r_A r_B} \, dV, \quad (11.89)$$

where $\mathbf{h}_{\text{eff}}^A$ and $\mathbf{h}_{\text{eff}}^B$ are the respective effective lengths of the antennas and r_A and r_B are their distances from the integration point. Noting Equation 11.27 for the mutual impedance Z_{AB} between the antennas in a uniform medium, we obtain that the additional mutual impedance due to the anomaly satisfies

$$\frac{\delta Z_{AB}}{Z_{AB}} = \frac{\beta^2 r_{AB}}{4\pi} \exp(2j\beta r_{AB}) \int_{V^+} \frac{\delta\epsilon}{\epsilon} \frac{\exp[-j\beta(r_A + r_B)]}{r_A r_B} \, dV, \quad (11.90)$$

where r_{AB} is the distance between the antennas. In the case of a scatterer with dimensions very much smaller than a wavelength, this simplifies to

$$\frac{\delta Z_{AB}}{Z_{AB}} = \frac{\beta^2 r_{AB} V^+}{4\pi r_A r_B} \frac{\delta\epsilon}{\epsilon} \exp[j\beta(2r_{AB} - r_A - r_B)], \quad (11.91)$$

where V^+ is the volume of the scatterer.

Under normal conditions, the atmosphere exhibits a significant amount of turbulence and this can result in significant permittivity variations. Because of this, even if there is no line of sight, Equation 11.90 implies that there can still be significant propagation. At VHF frequencies and above, scatter from atmospheric turbulence can provide an effective means of propagation over the horizon. Even in line-of-sight communication, scatter mechanisms can still have a significant effect. As shown in Figure 11.21, there will be scatter signal paths besides the direct path and this will cause a spread in transmission delay. The additional signals can also interfere with the main signal and hence cause fluctuations in signal level when there is relative motion between the

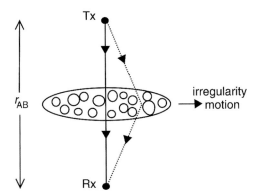

Figure 11.21   Scintillation caused by permittivity irregularity.

irregularity and the line of sight. This phenomenon, known as *scintillation*, is most evident in satellite communications and radio astronomical observations (the twinkle of stars). Permittivity irregularity can be caused by fluctuations in the ionosphere as well as the atmosphere. In addition, weather phenomena (such as rain, hail and snow) can also cause fluctuations in atmospheric permittivity and these can greatly affect propagation at frequencies in the high gigahertz range. The resulting scatter can cause signal loss that will be quite severe at such frequencies.

EXERCISES

(1) Two identical dipoles are placed a distance R apart at height H above a perfectly conducting ground. Both antennas have effective length $h_{\text{eff}} = h \cos \theta$, where θ is the angle from the horizontal. Find an expression for the voltage V induced in one antenna when a current I flows in the other. Investigate the variation of received voltage with H and R.

(2) Let two radio stations have vertically polarised antennas (free space gains G_A and G_B towards each other) that are at heights H_A and H_B above the ground. If the stations are separated by a distance R such that $R \gg H_A$ and $R \gg H_B$, show that the modified Friss equation has the form

$$P_R = P_B G_A G_B \left(\frac{\lambda}{4\pi R}\right)^2 \left|1 + R_V \exp\left(-2j\frac{\beta H_A H_B}{R}\right)\right|^2$$

and obtain an expression, valid in the limit of low ground impedance, for reflection coefficient R_V in terms of H_A, H_B and R.

(3) Two 500 MHz stations are located a distance 500 m apart with vertical dipole antennas that are 5 m above the ground ($\sigma = 0.1 \, \text{S/m}$ and $\epsilon_r = 10$). Calculate the loss of power in transmission between the two stations. If a 10 m high screen is placed at 25 m from one of the stations, calculate the additional loss.

(4) Write a computer program to calculate the range D at which radiation, launched at angle θ_0 from the vertical, lands after propagation via an ionospheric F layer (Equation 11.72). Use this program to verify that there is a region, that includes the source, within which no radiation lands. Noting that energy propagates parallel to a ray path, extend your program to allow estimation of the power loss during propagation. The loss is equal to the area of a ray tube, scaled on that at distance $\lambda/2\pi$ from the source (note that the rays also spread out in azimuth from the source and this gives additional loss to that arising from vertical spreading).

(5) An AM broadcast station operates on a frequency of 1 MHz and is expected to cover an area of radius 100 km around the chosen transmitter site. (The ground has conductivity 0.01 S/m and relative permittivity 10.) If the transmitter uses a quarter-wave monopole antenna (effective length $\approx 0.16\lambda$), calculate the transmitter power that is required to ensure a field of at least $100\,\mu$ V/m at all points in the coverage region.

SOURCES

Blaunstein, N. 2000. *Radio Propagation in Cellular Networks*. Norwood, MA: Artech House.

Budden, K. G. 1988. *The Propagation of Radio Waves*. Cambridge University Press.

Collin, R. E. 1985. *Antennas and Radio Wave Propagation*. New York: McGraw-Hill.

Davies, K. 1990. *Ionospheric Radio*. IEE Electromagnetic Waves series, Volume 31. London: Peter Peregrinus.

Felsen, L. B. and Marcuvitz, N. 1994. *Radiation and Scattering of Waves*. IEEE Press and Oxford University Press.

Hall, G. J. (ed.). 1988. *The ARRL Antenna Book*. Newark, CT: American Radio Relay League.

Jones, D. S. 1999. *Methods in Electromagnetic Wave Propagation* (2nd edn). IEEE/OUP series in Electromagnetic Wave Theory. Oxford University Press.

Monteath, G. D. 1973. *Applications of the Electromagnetic Reciprocity Principle*. Oxford: Pergamon Press.

Saunders, S. R. 1999. *Antennas and Propagation for Wireless Communication Systems*. Chichester: John Wiley.

Smith, A. A. 1998. *Radio Frequency Principles and Applications*. Piscataway, NJ: IEEE Press.

Wait, J. R. 1996. *Electromagnetic Waves in Stratified Media*. IEEE/OUP series in Electromagnetic Wave Theory. Oxford University Press.

Yeh, K. C. and Liu, C. H. 1972. *Theory of Ionospheric Waves*. New York: Academic Press.

12 Digital techniques in radio

The advent of high speed computers has opened up a whole new range of possibilities for radio. If the RF signal can be adequately represented by a series of samples (at a rate that a computer can handle), standard operations such as mixing, filtering, signal synthesis and demodulation can all be handled as mathematical operations within the computer. Constructing systems that can handle the complex signal processing required by spread spectrum communications, radar and other more exotic RF systems is merely an exercise in computer programming. Since most of the processing will be done inside a computer, we have what is commonly termed *software radio*. Such radios can be extremely flexible and be instantaneously reconfigured to handle new forms of modulation and/or tasking. All we need is a suitable *analogue-to-digital converter* (ADC) to interface to the incoming analogue signal and a suitable *digital-to-analogue converter* (DAC) to produce the outgoing analogue signal. Obviously, any realistic implementation of software radio will involve many constraints, and issues such as sampling rate and quantisation error will need to be addressed. The following chapter introduces the basic ideas of digital RF techniques and their limitations.

12.1 The processing of digitised signals

We have already noted the utility of studying RF systems in terms of the complex signal $\exp(j2\pi ft)$. Since $\cos(\theta) = \frac{1}{2}[\exp(j\theta) + \exp(-j\theta)]$, it is clear that the real signal $s(t) = S\cos(2\pi ft)$ will contain, in equal parts, contributions from frequencies f and $-f$. By means of its Fourier transform $\hat{s}(f) = \int_{-\infty}^{\infty} \exp(-j2\pi ft)s(t)\,dt$, a general signal $s(t)$ can also be represented as a linear combination of complex signals of the form $\exp(j2\pi ft)$ (note that $s(t) = \int_{-\infty}^{\infty} \exp(j2\pi ft)\hat{s}(f)\,df$). In the case of a real signal, the positive frequency components will be a mirror image of the negative frequency components (see Figure 12.1). If the positive frequency component has real part $s^I(t)$ and imaginary part $s^Q(t)$, the negative frequency component will have the same real part $s^I(t)$ and imaginary part $-s^Q(t)$. Quantity s^I is known as the *in-phase* component and s^Q as the *quadrature* component (note that the original real signal will have magnitude $2s^I$).

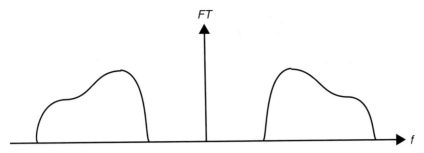

Figure 12.1   The frequency content of a real signal.

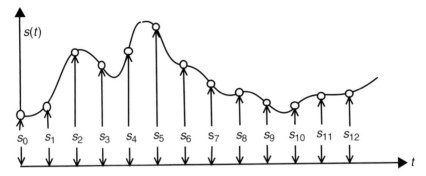

Figure 12.2   The sampling of an analogue signal.

For a computer to process an incoming analogue RF signal, it will need to derive a sequence of discrete samples that give an adequate representation of the signal. Consider an input signal $s(t)$ that has been sampled (see Figure 12.2) at a rate F for a finite number N of samples (there is an interval $T = 1/F$ between samples). We start with the relationship

$$\sum_{n=0}^{N-1} \exp\left(j\frac{2\pi n}{N}(l-k)\right) = \frac{1 - \exp(j2\pi(l-k))}{1 - \exp(j\frac{2\pi}{N}(l-k))} \tag{12.1}$$

for which the right-hand side takes the value N when $l = k$ and is zero for other integral combinations of l and k. If we denote the N signal samples by $s_0, s_1, s_2, \ldots, s_{N-1}$ and take a weighted sum with respect to the above expression, we obtain that

$$\frac{1}{N} \sum_{k=0}^{N-1} s_k \sum_{n=0}^{N-1} \exp\left(j\frac{2\pi n}{N}(l-k)\right) = s_l \tag{12.2}$$

and rearranging

$$\sum_{n=0}^{N-1} \left[\frac{1}{N} \sum_{k=0}^{N-1} s_k \exp\left(-j\frac{2\pi kn}{N}\right) \right] \exp\left(j\frac{2\pi ln}{N}\right) = s_l. \tag{12.3}$$

If we define

$$a_n = \sum_{k=0}^{N-1} s_k \exp\left(-j\frac{2\pi kn}{N}\right) \qquad (12.4)$$

we can approximate the original signal by

$$s(t) \approx \frac{1}{N} \sum_{n=0}^{N-1} a_n \exp\left(j2\pi n \Delta F t\right) \qquad (12.5)$$

for which there will be strict equality at the sample times of $0, T, 2T, 3T, \ldots,$ $(N-1)T$ (note that $\Delta F = F/N$). The coefficients $a_0, a_1, a_2, a_3, \ldots, a_{N-1}$ represent the content at frequencies $0, \Delta F, 2\Delta F, 3\Delta F, \ldots, (N-1)\Delta F$, respectively, and to-gether constitute what is commonly known as the *discrete Fourier transform* (DFT). It is important to appreciate some of the limitations of the discrete Fourier transform. Consider the complex signal $s(t) = \exp(j2\pi f t)$, where f is its frequency, this will have the DFT

$$\begin{aligned}
a_n &= \sum_{k=0}^{N-1} \exp\left(-j2\pi k\frac{n}{N}\right) \exp\left(j2\pi k\frac{f}{F}\right) \\
&= \frac{1 - \exp\left[-j2\pi N\left(\frac{n}{N} - \frac{f}{F}\right)\right]}{1 - \exp\left[-j2\pi\left(\frac{n}{N} - \frac{f}{F}\right)\right]} \\
&= \exp\left[-j\pi(N-1)\left(\frac{n}{N} - \frac{f}{F}\right)\right] \frac{\sin\left[\pi N\left(\frac{n}{N} - \frac{f}{F}\right)\right]}{\sin\left[\pi\left(\frac{n}{N} - \frac{f}{F}\right)\right]}.
\end{aligned} \qquad (12.6)$$

It will be noted that, for frequencies with $f = m\Delta F$ (m is an integer), we will have $a_m = 1$ and $a_n = 0$ when $n \neq m$. Consequently, the DFT picks out the correct spectral content. For frequencies f other than integral multiples of ΔF, the DFT will spread the energy around all the coefficients (this is known as *spectral leakage*). The major contributions, however, will be confined to coefficients associated with frequencies close to f. It is clear that ΔF (and hence the number of samples at a given sample rate) sets the lower limit to the frequency resolution of the DFT.

The question arises as to the behaviour of the DFT for frequencies outside the range of 0 to F. On noting that $\exp(j2\pi) = 1$, it is clear from the definition of the DFT that $s(t) = \exp[j2\pi(f + mF)t]$ will have the same DFT as $s(t) = \exp(j2\pi f t)$ when m is an integer. This property is known as *aliasing* and means that a DFT has the potential to *fold* the higher frequency content of a signal onto its lower frequency content. Figure 12.3 shows three signals with distinctly different frequency content, as illustrated by their Fourier transform (FT), but with the same DFT. In Figure 12.4, the signal has bandwidth greater than F and hence the DFT has folded the negative frequency content onto the positive frequency content. Clearly, unless the bandwidth of the signal is limited, the

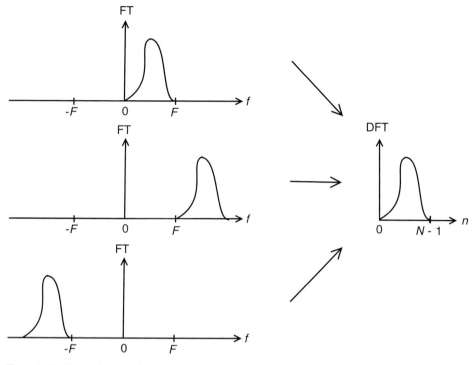

Figure 12.3   Three signals with identical DFT due to aliasing.

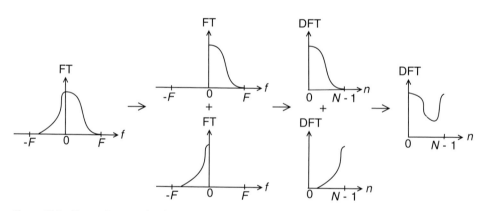

Figure 12.4   Spectral contamination caused by aliasing.

DFT will not be able to resolve the frequency content. For discrete sampling to be of any utility, it is essential that the signals are band limited. Now consider the case of real signals.

The simplest example will be a cosine signal $s(t) = \cos(2\pi f t) = \frac{1}{2}[\exp(j2\pi f t) + \exp(-j2\pi f t)]$. If we assume that $f = m\Delta F$, where m is an integer, this will only have two non-zero DFT coefficients, $a_m = \frac{1}{2}$ and $a_{N-m} = \frac{1}{2}$. Note, however, that the

DFT will only provide a unique specification of the cosine when $f \leq \Delta F/2$ since, if $f > \Delta F/2$, we will have no way of knowing which of a_m and a_{N-m} have been produced by positive and negative frequencies, respectively. This leads us to the *Nyquist sampling theorem*, a result that states:

The sample rate F will need to be at least twice the maximum frequency present if aliasing is to be avoided.

Before sampling a signal, it is normal to use an anti-aliasing filter in order to remove signal components at frequencies above $F/2$.

An alternative way of looking at finite discrete sampling is to regard it as the infinite discrete sampling of an alternative signal that is zero outside the original sampling interval. Clearly, resolution will no longer be a problem (we have an infinite number of samples). The alternative signal, however, will have abrupt changes at the transitions to zero state and will therefore have some very high frequency content (a step function has components on all frequencies). Some of this higher frequency content will be folded back onto lower frequencies by the DFT and this will be the cause of the spectral leakage we have already discussed. The solution to the problem is to introduce a process known as *windowing*. In windowing, we make the beginning and end samples smoothly transition to zero and hence remove the abrupt changes that cause the artificial high frequency content. The downside is that we have reduced the number of true samples and hence reduced the resolution of the DFT.

A sampled signal can be band-pass filtered (filtered with lower and upper band edges $P\Delta F$ and $Q\Delta F$, respectively), by reconstructing the sampled sequence from the DFT coefficients from a_P and a_Q. That is, we calculate the processed samples according to

$$s_l^{\mathrm{BPF}} = \frac{1}{N} \sum_{n=P}^{Q} a_n \exp\left(j2\pi \frac{ln}{N} \right). \tag{12.7}$$

The coefficients a_n can be replaced using Equation 12.4 to yield

$$s_l^{\mathrm{BPF}} = \sum_{k=0}^{N-1} s_k\, c_{k-l} \tag{12.8}$$

where

$$c_m = \frac{1}{N} \sum_{n=P}^{Q} \exp\left(-j2\pi \frac{mn}{N} \right). \tag{12.9}$$

This suggests that band-passed samples can be formed as linear combinations of the original samples. Indeed, Equation 12.9 suggest that an infinite band-passed sequence $(s_0^{\mathrm{BPF}}, s_1^{\mathrm{BPF}}, s_2^{\mathrm{BPF}}, s_3^{\mathrm{BPF}}, \ldots)$ can be formed from an infinite sequence of samples $(s_0, s_1, s_2, s_3, \ldots)$ by band-pass filtering successive blocks of N data samples. This

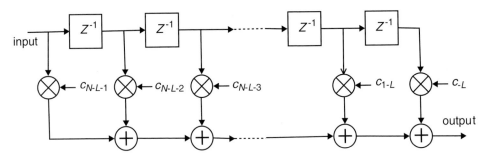

Figure 12.5   A finite length discrete signal filtering system.

implies a filtered sequence that is generated according to

$$s_M^{\text{BPF}} = \sum_{k=-L}^{N-1-L} s_{k+M}\, c_k \tag{12.10}$$

for suitably chosen L (note that the output sequence will be delayed by $N - 1 - L$ samples). In circuit terms, the above filtering procedure can be represented as shown in Figure 12.5. A stream of signal samples enters from the left and a continuous stream of processed samples exits to the right. As each sample enters, the rest are shifted by one to the right. Notation Z^{-1} indicates a decrement of one in the sequence index, $+$ represents the sum of sequence elements and the $\times$ elements represent multiplication by a filter coefficient. The above filter works fine for signal content at frequencies that are integral multiples of ΔF. Unfortunately, for other frequencies, spectral leakage will lead to considerable content at frequencies outside the range $P\Delta F$ to $Q\Delta F$. To overcome this, the end filter coefficients will need to include some tapering in order to window the data. In this case, it is appropriate to choose $L \approx N/2$.

12.2   Analogue-to-digital conversion

The simplest analogue-to-digital converter (ADC) is that which simply detects whether a pulse is present (i.e., the threshold detection that is required in certain types of digital communication). We could implement this as a comparator that is based on an operational amplifier (see Figure 12.6). When the input voltage is above a certain threshold, its level is raised to that of the supply. Below the threshold, however, the output will be forced to the ground voltage. Software radio, however, will require far more resolution in amplitude than this one-bit converter can provide. Figure 12.7 shows a simple *flash converter* with two-bit accuracy (i.e., the converter can detect four amplitude levels). The input of the converter has a *sample and hold* system to ensure that the samples have a constant amplitude during the acquisition period. The digitisation takes place through a multi-threshold comparator and logic circuits are used to encode the output as a two-bit binary number. A two-bit converter is still too crude

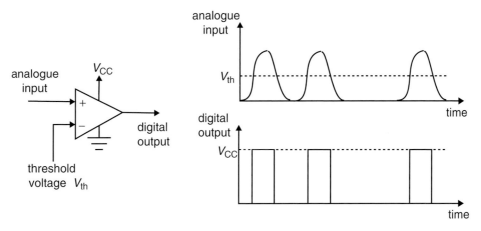

Figure 12.6   A simple one-bit analogue-to-digital converter.

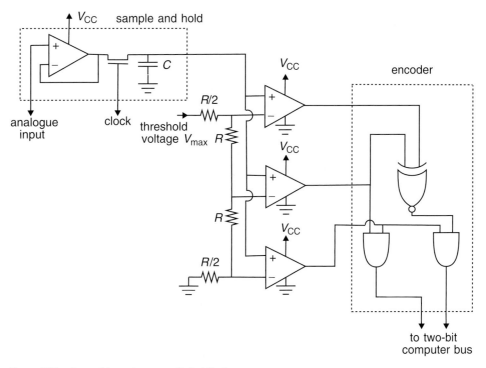

Figure 12.7   A two-bit analogue-to-digital flash converter.

for software radio (a 14-bit converter is far more usual), but demonstrates the general principle (Figure 12.8).

It is clear that, with only a finite number of fixed levels for voltage representation, the digitisation process will lead to some inaccuracy. This inaccuracy manifests itself as *quantisation noise* and will, for levels with spacing Δ, have extremes of $\pm\Delta/2$ in the

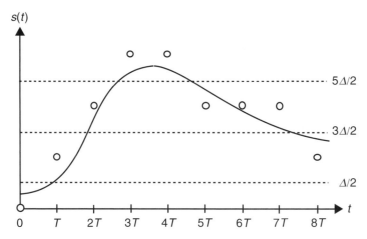

Figure 12.8 Digitisation of a general signal.

sampling interval T. We assume that the quantisation voltage error v_{qn} varies linearly between these extremes and so its mean square will be given by

$$\overline{v_{qn}^2} = \frac{1}{T}\int_0^T v_{qn}^2 \, dt = \frac{1}{T}\int_0^T \left(\frac{\Delta}{2} - \frac{\Delta}{2}\frac{t}{T}\right)^2 dt = \frac{\Delta^2}{12} = \frac{V_{max}^2 2^{-2b}}{12}, \qquad (12.11)$$

where b is the bit number resolution of the ADC and V_{max} is the voltage range. The power of the quantisation noise will be

$$N_{qn} = \frac{V_{max}^2 2^{-2b}}{12R}, \qquad (12.12)$$

where R is the input resistance. Since V_{max} will be the peak-to-peak voltage of the largest sine wave that the converter can handle, the largest possible SNR of the converter will be

$$\mathrm{SNR} = \frac{(V_{max}/2)^2}{2R}\bigg/ N_{qn} = \frac{3}{2}2^{2b}$$
$$= 6.02b + 1.75 \text{ dB} \qquad (12.13)$$

which is the dynamic range of the converter. If we assume the quantisation noise to be spread uniformly across the *Nyquist bandwidth* $(F/2)$, the SNR will increase by factor $F/2B$ if the converter output is processed through a filter with bandwidth B. If the processing is through a DFT of length N, this will have a resolution of $\Delta F = F/N$ and hence an SNR of $6.02b + 1.75 + 10\log_{10}(N/2)$ dB in each resolution cell. Quantity $10\log_{10}(N/2)$ is known as the processing gain. Unfortunately, the assumption of uniformly distributed quantisation noise is only a crude approximation unless the converter input consists of a considerable number of independent signals. In the case of an input consisting of relatively few signals, the quantisation noise can become highly correlated and cause a considerable number of high amplitude *spurs*. The solution is to add some noise at the input to the ADC (a process known as *dithering*) in order to

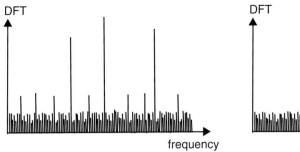

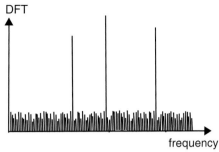

(a) Quantisation noise with spurs   (b) Noise with spurs removed by dither

Figure 12.9   Sampled signals with quantisation noise.

assist the decorrelation of the quantisation noise and hence reduce the spurs. Normally, such noise is added into a section of the Nyquist bandwidth that is to be removed by subsequent processing (Figure 12.9).

Another source of ADC noise is the fluctuation in sample interval that arises due to imperfections in a realistic clock signal. This *aperture jitter*, as it is known, will cause distortions to the input signal that translate into additional noise that increases with frequency and which, for high resolution converters, can be dominant. In this case

$$\text{SNR} = 20 \log_{10} \left(\frac{1}{2\pi f \, \Delta \tau} \right) \text{dB}, \qquad (12.14)$$

where $\Delta \tau$ is the rms value of the sample interval fluctuations and f is the frequency.

Once converted into digital form, the signal can be processed by a computer or perhaps some programmable logic. Often, however, we will need to convert the processed signal back into analogue form. To achieve this, we need a digital-to-analogue converter (DAC) and Figure 12.10 shows a simple converter that is based upon a voltage summing circuit.

12.3 Digital receivers

Consider the simple direct conversion receiver shown in Figure 12.11. In this design, the analogue-to-digital conversion is performed at baseband. For the cosine RF signal $s^{\text{RF}} = S^{\text{RF}} \cos(2\pi f_{\text{RF}} t)$, and local oscillator signal $s^{\text{LO}} = S^{\text{LO}} \cos(2\pi f_{\text{LO}} t)$, the in-phase signal (labelled I) leaving the mixers will be

$$
\begin{aligned}
s^{\text{I}}(t) &= S^{\text{RF}} S^{\text{LO}} \cos(2\pi f_{\text{LO}} t) \cos(2\pi f_{\text{RF}} t) \\
&= \frac{S^{\text{RF}} S^{\text{LO}}}{2} \{\cos[2\pi (f_{\text{LO}} - f_{\text{RF}})t] + \cos[2\pi (f_{\text{LO}} + f_{\text{RF}})t]\} \qquad (12.15)
\end{aligned}
$$

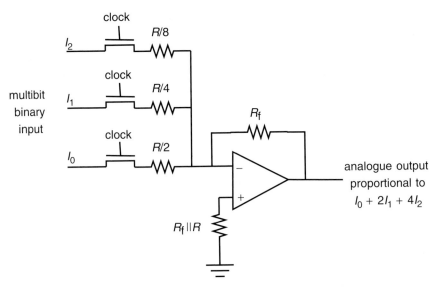

Figure 12.10　Simple digital-to-analogue converter.

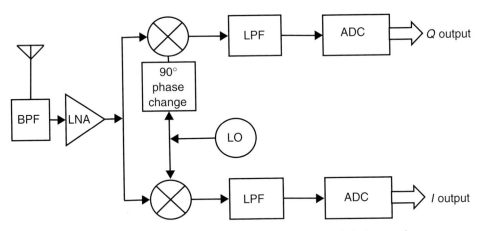

Figure 12.11　A direct conversion receiver with baseband analogue-to-digital conversion.

and the quadrature signal (labelled Q)

$$s^Q(t) = S^{RF} S^{LO} \sin(2\pi f_{LO} t) \cos(2\pi f_{RF} t)$$

$$= \frac{S^{RF} S^{LO}}{2} \{\sin[2\pi(f_{LO} - f_{RF})t] + \sin[2\pi(f_{LO} + f_{RF})t]\}. \qquad (12.16)$$

The low-pass filters at the mixer outputs will remove the high frequency components. Consequently, the effect of the receiver is to produce two copies of the input RF signal, both downward shifted by a frequency by f_{LO}. One copy will be in phase with the original signal and the other in quadrature (90° out of phase).

Let s_n^I represent an element of the in-phase ADC output sequence and s_n^Q an element of the quadrature output sequence. In the case of amplitude demodulation, the corresponding element of the demodulated output sequence can be calculated according to $S_n^{demod} = (|s_n^I|^2 + |s_n^Q|^2)^{1/2}$ and, in the case of phase modulation, according to $S_n^{demod} = \arctan(s_n^Q/s_n^I)$. Frequency demodulation is a little more difficult since it involves differentiating the phase demodulated output. In the discrete domain, however, this can be achieved with the difference sequence $S_n^{demod} = (1/T)[\arctan(s_n^Q/s_n^I) - \arctan(s_{n-1}^Q/s_{n-1}^I)]$, where T is the time between samples.

The demodulation of single sideband (SSB) signals is a lot more involved. In the case of lower sideband (LSB), the signal will reside below the carrier frequency and, in the case of upper sideband (USB), above this frequency. Let the quadrature output be further processed by a filter that shifts the phase by $-90°$ for positive frequencies and by $90°$ for negative frequencies (a *Hilbert transform* filter). In the case of our cosine RF signal, this will produce output of the form

$$s^H(t) = -\frac{S^{RF}S^{LO}}{2} \cos[2\pi(f_{LO} - f_{RF})t] \quad f_{RF} > f_{LO}$$

$$= \frac{S^{RF}S^{LO}}{2} \cos[2\pi(f_{LO} - f_{RF})t] \quad f_{RF} < f_{LO}, \quad (12.17)$$

where it will be noted that there is a distinct difference between the RF signals above and below the carrier frequency f_{LO}. The Hilbert transform will essentially multiply positive frequency content by $-j$ and negative frequency content by j. It is clear that, in order to demodulate SSB signals, we will need f_{LO} equal to the carrier frequency. We form output sequence $S_n^{demod} = s_n^I + s_n^H$ for LSB and $S_n^{demod} = s_n^I - s_n^H$ for USB (note that the terms s_n^H represent the quadrature samples s_n^Q after processing by a discrete version of the Hilbert transform filter).

The advent of high speed, high dynamic range, analogue-to-digital converters has led to the possibility of direct digitising RF receivers. In such a receiver, the mixer function is performed, after digitisation, by multiplying each term s_n of the digitised RF signal by the corresponding term of a digitised local oscillator sequence. A typical direct digitising receiver is shown in Figure 12.12. It will be noted that the filters on the I and

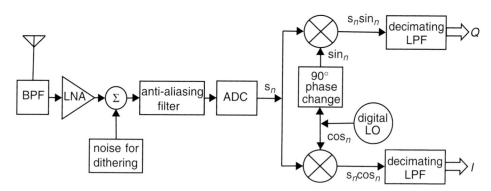

Figure 12.12 A direct digitising receiver.

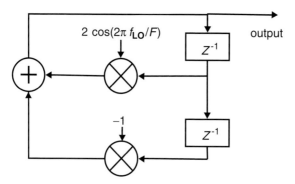

Figure 12.13   A digital oscillator.

Q channels are designated as *decimating filters*. Since these are low-pass filters, their output does not need to be produced at the same rate as the original sampling. We can throw away all of the output except for every jth sample (the *decimation* process), just so long as the new rate is above the Nyquist rate corresponding to the filter bandwidth (a rate of twice the bandwidth). This reduced output rate can significantly lower the computational demand of demodulation and subsequent process. One important issue is the means by which the discrete samples corresponding to the local oscillator are generated. This sequence can be formed by means of the direct digital synthesis procedure to be discussed in the next section. It is also possible, however, to produce the necessary samples by means of the digital oscillator shown in Figure 12.13 (note that the sequence is advanced by one for each sample period). If we start the sine sequence with $\sin_0 = 0$ and $\sin_1 = A \sin(2\pi f_{LO}/F)$, we can form the rest of the sequence according to $\sin_n = 2 \cos(2\pi f_{LO}/F)\sin_{n-1} - \sin_{n-2}$, where f_{LO} is the required oscillator frequency and F the sample rate. In a similar fashion, a cosine sequence can be constructed according to $\cos_0 = 1$ and $\cos_1 = A \cos(2\pi f_{LO}/F)$ with the rest of the sequence generated according to $\cos_n = 2 \cos(2\pi f_{LO}/F)\cos_{n-1} - \cos_{n-2}$.

At present, ADCs with suitable dynamic range are only available for the lower VHF frequencies. For frequencies above about 50 MHz, it is necessary to down convert and employ a digital receiver at an intermediate frequency. Although analogue down conversion is necessary for frequencies above about 500 MHz, it is possible to use *under sampling* techniques at frequencies below this. We have already noted that sampling a signal at frequency F will cause components at frequencies above $F/2$ to fold onto lower frequencies. If these lower frequency are unoccupied, we essentially have down conversion without the use of mixers. For this approach to be effective, we will need the band-pass filter at the receiver input to have a bandwidth that is less than the Nyquist bandwidth of the ADC. Furthermore, the sample and hold circuits in the ADC will need to have a response time that is consistent with the frequencies of the input signal.

12.4   Direct digital synthesis

The advent of high speed digital electronics has also opened up new possibilities for high quality frequency synthesis. Synthesisers based on phase locked loops constitute the core of many modern RF systems, but can be limited by their phase noise and slow settling times (problems caused by the use of a VCO). An approach to synthesis that reduces these problems is known as direct digital synthesis (DDS). Essentially, the approach uses the stored samples (a very large number) from one period of a perfect sine wave and outputs these repeatedly to a DAC. Samples are always sent to the converter at the same rate, but the output frequency is varied by selecting which of the samples is sent. Figure 12.14 shows a possible realisation of such a system. The samples are stored sequentially in a memory that outputs a sample from a programmable address when it receives a clock signal. The address is decided by the output of an accumulator circuit. If N is the number of stored samples, and the accumulator increments the address by n on each clock signal (period T), the output signal will exhibit the frequency $f = n/NT$. The address accumulator needs to have a programmable increment of, at most, $N/2$ (maximum frequency $f_{max} = 1/2T$) and needs to reset to zero when the limit N is reached. Due to the approximate form of the DAC output, it will contain harmonics and these must be removed by a low-pass filter (this filter will need to remove frequencies above f_{max}). The quality of the output will also be affected by the noise created through the limited bit resolution of both the memory and the DAC. If M is the number of bits of resolution, the output will have an SNR of approximately $6M$ dB.

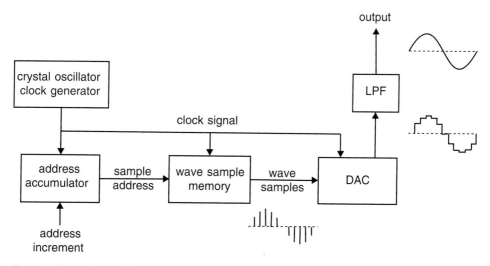

Figure 12.14   A direct digital synthesiser.

At present, the available DAC and memory technology only allows DDS systems to be used up to frequencies of around 200 MHz, but this limit is being improved at a rapid rate. DDS systems are ideal for software radio since they can be reconfigured almost instantaneously.

EXERCISES

(1) The signal

$$s(t) = \cos\left(2\pi\frac{t}{T}\right) + \cos\left(6\pi\frac{t}{T}\right) + \cos\left(10\pi\frac{t}{T}\right) \qquad (12.18)$$

is sampled at rate F. Discuss the DFT of this signal for sample rates $F = 1/T$, $F = 2/T$, $F = 6/T$ and $F = 10/T$. What is the influence of the number of samples?

(2) A digital spectral surveillance receiver is required to cover a 10 MHz frequency band with resolution 1 kHz. The system is to handle a maximum signal strength of 1 dBm and is fed by an antenna with noise temperature $300\,^\circ$K. Calculate the bit accuracy of an ADC required for the system to be externally noise limited.

(3) An ADC has a sample rate of 60 mega-samples per second and is required to form an under-sampling receiver for operation within the 95 to 115 MHz band. The receiver must handle a maximum signal strength of 1 dBm and is fed by an antenna with noise temperature $280\,^\circ$ K. If the signals are demodulated in 10 kHz bandwidths, calculate the bit accuracy of the ADC for the system to be externally noise limited. Calculate the order of the Butterworth filter that is required to ensure that the effect of aliasing is hidden by the noise.

SOURCES

Proakis, J. G. and Manolakis, D. G. 1996. *Digital Signal Processing*. Englewood Cliffs, NJ: Prentice-Hall.

Rhode, U. L., Whitaker, J. and Bucher, T. T. N. 1996. *Communication Receivers* (2nd edn). New York: McGraw-Hill.

Straw, R. Dean (ed.). 1999. *The ARRL Handbook* (77th edn). Newark, CT: American Radio Relay League.

Sabin, W. E. and Schoenike, E. O. (eds.). 1998. *HF Radio Systems and Circuits*. Atlanta, GA: Noble Publishing Corporation.

Reed, J. H. *Software Radio: A Modern Approach to Radio Engineering*. Englewood Cliffs, NJ: Prentice-Hall.

Index

active load, 66, 81
aliasing, 295
AM demodulator, 127
AM transmitter, 95
amplifier noise, 82
amplitude modulation (AM), 3, 95
amplitude shift keying (ASK), 19
angle modulation, 101
antenna efficiency, 232
antenna feed, 243
antenna interaction, 250
antenna temperature, 10
antennas, 228
aperture antennas, 260
array antennas, 247
automatic gain control (AGC), 103
axial mode helical antenna, 259

balanced circuit, 184
balanced diode mixer, 88
balanced feed, 244
BALUN, 184, 245
bandwidth, 3, 13, 31, 69
Barkhausen criterion, 109
baseband signal, 95
beamwidth, 251
Beverage antenna, 258
biconical dipole, 235
bipolar junction transistor (BJT), 50
BJT bias, 51
BJT Colpitts oscillator, 109
BJT mixer, 90
BJT model, 56
blocking capacitor, 54
bounded source, 224
bow tie dipole, 235
broadband monopole, 243
broadside, 250
bypass capacitor, 54

capacitive transformer, 37
capacitor, 44
carrier wave, 95
cascode amplifier, 62, 78
cellular radio, 23
channel capacity, 20
characteristic impedance, 134

charge pump, 125
choke dipole, 245
circular polarisation, 209
circular waveguide, 222
circulator, 164
Clapp oscillator, 112
class A amplifier, 171
class AB amplifier, 175
class B amplifier, 173
class C amplifier, 176
class E amplifier, 178
coaxial feed, 244, 245
coaxial transmission line, 140
common-base amplifier, 60
common-collector amplifier, 62
common-emitter amplifier, 54
compression, 15
conducting material, 205
conduction angle, 171
continuity equation, 208
conversion gain, 88
coupling coefficient, 35
current gain, 51
current mirror, 65
current source, 65, 81
cut-off frequency, 56, 220

damping ratio, 29
decimation, 304
decoupling, 71
desensitisation, 16
detectors, 86
dielectric resonator, 223
dielectric waveguide, 223
differential amplifier, 64
diffraction, 275
digital demodulation, 303
digital filter, 297
digital modulation, 19
digital receivers, 301
digital signal processing, 293
diode, 49
diode mixer, 86
diode ring mixer, 88
dipole antenna, 3, 228
direct conversion receiver, 105, 301
direct digital synthesis (DDS), 305

directional coupler, 160, 162
directivity, 5, 236
discone antenna, 241
discrete Fourier transform (DFT), 293
dispersion, 210
dither, 300
Doppler shift, 25
double balanced mixer, 94
dual-gate MOSFET, 80
dual-gate MOSFET mixer, 90
ducting, 281

eddy currents, 46
effective length, 6, 235
effective permittivity, 140, 210
efficiency, 173
electric dipole, 233
electromagnetic waves, 204
emitter degeneration, 82
emitter follower amplifier, 62
endfire, 250
envelope detector, 95
extended Rumsey principle, 254

feedback, 67, 108
ferrite core, 44
FET bias, 73
FET Colpitts oscillator, 111
FET mixer, 89
field effect transistor (FET), 71
flash A to D converter, 298
flicker noise, 9
flutter, 273
FM demodulation, 102
FM demodulator, 127
FM modulator, 101, 127
folded dipole, 233
Fraunhofer zone, 225
frequency modulation (FM), 3, 101
frequency shift keying (FSK), 19
frequency synthesiser, 127
Fresnel zone, 225
Friis equation, 8, 272

gain, 5, 237
gain control, 103
Gilbert cell mixer, 94
ground wave propagation, 286
group speed, 211
Gunn diode, 121

half-wave dipole, 230
harmonics, 15
Hartley oscillator, 112
helical monopole, 241
Helmholtz equation, 224
homodyne receiver, 105
horn antenna, 261
hybrid, 163

ideal dipole, 229
inductor, 44
instability, 156
integrated circuit (IC), 65
inter-modulation distortion (IMD), 16
ionospheric F layer, 281
ionospheric propagation, 281

JFET, 71

L-network matching, 38
limiting amplifier, 104
linear polarisation, 209
load line, 51
log periodic dipole array (LPDA), 254
Lorentz force, 205
Lorentz reciprocity theorem, 267
loss tangent, 141
lower sideband, 96

magnetic dipole, 233
magnetic vector potential, 224
matching circuits, 28
Maxwell's equations, 204
microstrip transmission line, 140
Miller capacitance, 59
Miller effect, 57, 59, 77
minimum detectable signal, 14, 16
mismatch efficiency, 247
mixers, 86
mobile communications, 273
modulation, 3
modulation index, 95
modulators, 86
monopole, 238
MOSFET, 71

natural frequency, 29
negative resistance, 121
network analyser, 161
nMOS, 71
noise, 9
noise factor, 11
noise figure, 11
noise temperature, 10
non-linear effects, 14
Nyquist bandwidth, 300
Nyquist sampling theorem, 297

oblique reflection, 215
Ohm's law, 205
ohmic loss, 46
ohmic resistance, 232
open-circuit line, 142
oscillators, 108

parallel plate capacitor, 46
parallel resonance, 33
parallel wire capacitor, 46
parasitic oscillation, 71
patch antenna, 261

perfect conductor, 205
permeability, 205
permittivity, 205
phase comparator, 123, 125
phase locked loop, 123
phase margin, 70
phase modulation (PM), 3, 101
phase noise, 117
phase path, 279
phase shift keying (PSK), 19
phase shift network, 99
phase speed, 211
phasing SSB demodulator, 99
phasing SSB generator, 97
physical optics, 260
π-network matching, 39
piezoelectric crystal, 119
plane wave, 207
PLL capture range, 124
PLL lock range, 124
PLL filter, 125
PM demodulation, 102
PM modulator, 101
pMOS, 71
polarisation, 7
polarisation efficiency, 272
polarisation match, 272
polarisation vector, 286
Poynting vector, 206, 231, 285
printed antenna, 263
propagation, 266
propagation constant, 135

quadrature phase shift keying (QPSK),
    20
quality factor (Q factor), 31
quantisation noise, 299
quarter-wave choke, 245
quarter-wave transformer, 142
quiescent state, 52

radar, 23
radar cross-section, 25
radar equation, 25
radiation loss, 232
radiation resistance, 232
radiation zone, 225
radiating near-field zone, 225
radio waves, 1
range ambiguity, 24
range resolution, 25
ray path, 279
reactive near-field zone, 225
reciprocal mixing, 119
reciprocity, 8, 266, 289
reflection, 272
reflection coefficient, 137, 217
reflections at an interface, 211
reflector antenna, 261
refraction, 277
refractive index, 278
resistor, 44

resonance, 28
resonant circuits, 28
resonant frequency, 28
Rollet factor, 157
Rumsey principle, 235

S parameter amplifier design, 153
S parameter measurement, 160
S parameters, 148
sample and hold, 298
sampling, 294
saturation current, 50
saturation region, 72
scatter propagation, 290
scattering, 287
scintillation, 291
selectivity, 13
self-resonance, 46
sensitivity, 13
series resonance, 28
shadow region, 275
Shannon–Hartley theorem, 20
short-circuit line, 142
short dipole, 232
shot noise, 9, 83
signal-to-clutter ratio (SCR), 26
signal-to-interference ratio (SIR), 23
signal-to-noise ratio (SNR), 11
single sideband (SSB) modulation, 97
skin depth, 214
skin effect, 46, 214
skip zone, 284
Smith chart, 146
Snell's law, 277, 282
software radio, 293
solenoid, 44
spectral leakage, 295
spectrum analyser, 17
spread spectrum, 21
spurious free dynamic range (SFDR), 16
stability, 70, 117, 156
stub matching, 143
superheterodyne receiver, 13, 99
synchronous demodulation, 96
synchronous demodulator, 99

T-network matching, 40
thermal noise, 9, 83
thermal voltage, 50
time harmonic, 30, 135, 208
toroidal inductor, 45
toroidal transformer, 45
transconductance, 76
transconductance mixer, 93
transducer gain, 155
transformer, 34
transmission line, 132
transmission line loss, 141
transmission line transformers, 181
transverse electric (TE) waves, 220
transverse electromagnetic (TEM) waves,
    209

transverse magnetic (TM) waves, 220
travelling wave antenna, 258
triode region, 71
tuned transformer, 36

unbalanced circuit, 184
unconditionally stable, 156
under-sampling, 304
upper sideband, 96

voltage controlled oscillator (VCO), 120, 123
voltage standing wave ratio (VSWR), 137

waveguides, 218
Wheeler result, 233
Wilkinson power divider, 161
windowing, 297

Yagi–Uda antenna, 257